JEFFREY DINGLE

PEREGRINE PRESS MARBLEHEAD, MA

ISBN 0-9637162-5-5

Fifth Edition

ESSENTIAL RADIO GUIDE is a registered trademark
of PEREGRINE PRESS LTD.

RADIOBASE ® database
provided by
THE CENTER For RADIO INFORMATION
19 Market Street
Cold Spring, NY 10516

Tel: 800.359.9898
Fax: 914.265.2715
www.the-cri.com
e-mail: info@the-cri.com

ESSENTIAL RADIO GUIDE was produced
on a Macintosh PowerBook ® 1400c/166.

For information, please contact:

PEREGRINE PRESS LTD.
P.O. Box 363
Marblehead, MA 01945

Tel: 781.639.8090
Fax: 781.639.6341
e-mail: info@webradionet.com

CONTENTS

- ACKNOWLEDGEMENTS -

Since I first embarked on what I refer to as my "ESSEN-TIAL RADIO" project, the research, development and marketing of the book has consumer thousands of hours. Although I have been responsible for, and at times obsessed with every minute detail of ESSENTIAL RADIO, there have been many people who have contributed to the venture as it has developed over the years - whether directly or indirectly - and collectively, they have played a significant role in making it all come together. I look forward to their continued participation in the development of ESSENTIAL RADIO.

I am particularly grateful to Bruce Jones who has provided ongoing advice and information over the past five years on numerous aspects of the design, printing and publishing industry. Bruce also designed the cover.

My sister Leslie Carrere designed PEREGRINE PRESS's logo. My attorney Greg White has been extremely helpful over the years, both in organizing all the legal aspects of the business, as well as playing the role of a consultant. Mike Squires has played a key role in the promotion and marketing of ESSENTIAL RADIO.

Several other people have also provided important advice, input and support to ESSENTIAL RADIO. These include Ralph Guild, Maurie and Scott Webster at THE CENTER FOR RADIO INFORMATION, my mother Libby Warner and George and Gladys Poor. And of course, Beau, Liza, Clare and Jeremy.

FORMATS

AC	-	Adult Contemporary
ALT RK	-	Adult Album Alternative Rock
BB	-	Big Band
CLS	-	Classical
CLS RK	-	Classic Rock
CTRY	-	Country
DVRS	-	Diverse
EDU	-	Education
ETH	-	Ethnic
EZ	-	Easy Listening
GOSP	-	Gospel
HOT AC	-	Hot Adult Contemporary
JZZ	-	Jazz
KIDS	-	Children
LGHT AC	-	Light Adult Contemporary
MOD RK	-	Modern Rock
NOS	-	Nostalgia
NPR	-	National Public Radio
NWS	-	News
OLDS	-	Oldies
PRI	-	Public Radio International
RLG	-	Religious
RLG AC	-	Contemporary Christian
ROCK	-	Album Oriented Rock (AOR)
70's RK	-	70's Rock
SPAN	-	Spanish
SPRTS	-	Sports
TLK	-	Talk
TOP 40	-	Top 40 / Contemporary Hit Radio (CHR)
URBAN	-	Urban Contemporary

"C" - COLLEGE STATION   •   "P" - PUBLIC STATION

SIGNAL STRENGTH

I	- Weak. Clear signal 10-15 miles from source.
III	- Medium. Clear signal 15-25 miles from source.
IIIII	- Strong. Clear signal 30-50 miles from source.
IIIIIIII	- Very Strong. Clear signal 50-80 miles from source.

NOTE: Reception of an FM signal is dictated by "line-of-sight" between the source and the receiver. Accordingly, local topography (mountains, buildings, etc.) can significantly influence FM reception. The above signal signal strength quantifications, which incorporate power output and tower height, are approximate and are based on ideal topographic conditions.

- FOREWORD -

The idea for *ESSENTIAL RADIO GUDIE* was first sown in my mind more than ten years ago as I sat in a rental car at an airport, working the Scan button on the radio hoping to land on a station that played something I wanted to listen to. Perhaps you know the feeling? A simple guide or directory that provided basic information on local radio stations would have solved my problem - except nothing like it existed - at least not that I could find.

Five years ago the idea surfaced once again. This time I decided to take a serious look at what it would take to turn my idea into a product. The end result was "*ESSENTIAL FM*", a directory that listed 5,000 FM stations in the United States first published in October 1993 (FM radio commands about 75% of the national listening audience). Six months later, *ESSENTIAL FM* was revised to include approximately 1,000 AM stations, and retitled "*ESSENTIAL RADIO GUDIE*". A second section cross referencing stations by frequency was also added to *ESSENTIAL RADIO GUDIE* as well as a listing of the flagship stations of the major professional sports teams. Five years, five editions and thousands of copies later, I am pleased to be publishing the Fifth Edition of *ESSENTIAL RADIO GUDIE*.

Being a pocket guide obviously forces one to make some compromises with respect to how much information is provided. Accordingly, *ESSENTIAL RADIO GUDIE* provides the radio listener with primary station information - Frequency, Call Letters, Signal Strength, and of course, Station Program Formats. The fifth edition of *ESSENTIAL RADIO GUDIE* lists approximately 4,700 stations, of which approximately 75% are FM and 25% are AM. 270 markets are covered.

One question I have been asked is why not list every station? A guide listing every station, which I may publish at a future date, would be nearly twice as large and more expensive, and become more of a reference book than a travel guide. With more than 4,500 stations covering nearly 300 markets, most listeners should find ample selections. My apologies to those stations who are not listed. Several other ideas on the drawing board - a Sports Radio guide, a Talk Radio guide, and *THE WEB RADIO NETWORK* to name a few - which I hope to pursue in the future.

Although the data base is revised on an ongoing basis (provided by The Center for Radio Information), no doubt there are stations listed in *ESSENTIAL RADIO GUDIE* that have either changed their call letters and/or program formats, moved to a new frequency or gone off the air - or are new on the air - since going to press.

A final note - if you're driving alone, be careful using *ESSENTIAL RADIO GUDIE*.

Jeffrey Dingle
October 1998 - Marblehead, MA

HOW TO USE
ESSENTIAL RADIO GUIDE

Using ESSENTIAL RADIO is basically self-explanatory. However, it is helpful to understand a few features regarding the page layout, how formats are organized and grouped and signal strength. Here are some of the ESSENTIAL RADIO's basic features.

STATION LISTINGS

SECTIONS A-G segment the United States into **Seven Geographic Regions**. Approximately **270 markets** are covered. Each section is comprised of **two parts** - a list of **stations by market** and a list that **cross references** each station by **frequency**.

For station **listings by market**, the following information is provided:

FREQUENCY • CALL LETTERS • SIGNAL STRENGTH • FORMAT.

COLLEGE or **PUBLIC** station are referenced with a "C" or "P" in the right hand margin.

For station **listings by frequency**, the following information is provided:

FREQUENCY • STATE • MARKET • PAGE NUMBER

Due to signal strength and local topography, clear reception (and in some cases station reception at all) will not always be found until one is close to the actual city. This is particularly true in mountainous regions. AM radio is also affected by atmospheric conditions, which can significantly impede signal reception. However, at night, a strong AM signal, in particular, those of Clear Channel can sometimes be picked up 500-750 miles from the actual source of transmission.

FORMAT GROUPS

There are **ELEVEN** basic **FORMAT GROUPS**. In cities where there are multiple stations classified within a single format group, all of these stations are grouped together. Format groups are organized in ascending order, by the primary format of the station, as follows:

1. ADULT CONTEMPORARY (POP)
2. TOP 40, CONTEMPORARY HIT RADIO
3. ROCK, CLASSIC ROCK, MODERN ROCK, ALTERNATIVE ROCK, SOFT ROCK
4. OLDIES
5. COUNTRY
6. SPANISH
7. CLASSICAL, JAZZ, BIG BAND
8. ALTERNATIVE, DIVERSE, NPR, PRI (PUBLIC & COLLEGE)
9. RELIGIOUS, ETHNIC, EDUCATION, GOSPEL, EASY LISTENING
10. URBAN CONTEMPORARY, RHYTHM & BLUES
11. NEWS, TALK, SPORTS

For example, Adult Contemporary stations will always be the first stations listed in a given city, Top 40 the second, etc.. News, Talk and Sports stations are always the last stations listed in a given city. Also note that many stations play more than one program format.

MAPS

Each section has a Map page at the beginning. The maps show the states listed in that section and list the individual markets covered and their page numbers.

SIGNAL STRENGTH

Every station includes an icon, **"lll"**, indicating the strength of that station's signal. With FM radio, signal propagation is a function of the stations power output (watts) and the height of the tower above average terrain (referred to as HAAT). **"l"** indicates a weak signal and/or low tower; **"lllllll"** indicates a very strong signal and a very tall tower. However, due to the nature of the FM signal, whose reception is primarily dictated by "line-of-sight" between the source and the receiver, the relative strength of a station's signal offers only a partial indicator of whether or not one will be able to pick up a given station.

Local topography will have a significant influence on reception of a radio station (as will the quality of one's receiver or car stereo) and this must be kept in mind when referencing FM signal strength. AM reception is also influenced by atmospheric conditions.

PROFESSIONAL SPORTS TEAMS

SECTION H lists the flagship stations of the major professional sports teams - MAJOR LEAGUE BASEBALL, NATIONAL FOOTBALL LEAGUE, NATIONAL BASKETBALL ASSOCIATION and NATIONAL HOCKEY LEAGUE. Spanish (and French) flagship stations are included where applicable and referenced with "SPAN" or "FREN" in the right margin.

800 NUMBERS

SECTION I lists the 800 numbers of the major AIRLINE, HOTEL and CAR RENTAL companies.

A

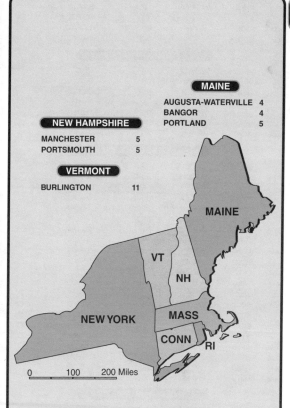

MAINE

AUGUSTA-WATERVILLE	4
BANGOR	4
PORTLAND	5

NEW HAMPSHIRE

MANCHESTER	5
PORTSMOUTH	5

VERMONT

BURLINGTON	11

NEW YORK

ALBANY-SCHENECT.	5
BINGHAMTON	6
BUFFALO	6
ELMIRA	7
ITHACA	7
LONG ISLAND	7
NEW YORK CITY	8
NEWBURGH-MIDDL	8
POUGHKEEPSIE	8
ROCHESTER	9
SYRACUSE	9
UTICA	10
WATERTOWN	10

MASSACHUSETTS

BOSTON	3
CAPE COD	4
SPRINGFIELD	4
WORCESTER	4

RHODE ISLAND

PROVIDENCE	10

CONNECTICUT

BRIDGEPORT	2
DANBURY	2
HARTFORD	2
NEW HAVEN	2
NEW LONDON	2
STAMFORD	2
WATERBURY	3

	FREQ	STATION	POWER	FORMAT	
	99.9	WEZN	IIIII	AC	
AM	1450	WCUM	I	SPAN	
	89.5	WPKN	III	DVRS	C
AM	1500	WFIF	IIIII	RLG AC	
AM	600	WICC	IIIII	NWS, TLK	

DANBURY, CT

	FREQ	STATION	POWER	FORMAT	
	98.3	WDAQ	III	AC	
	95.1	WRKI	IIIII	ROCK	
AM	850	WREF	IIIII	OLDS	
	88.5	WVOF	I	DVRS	C
	91.1	WSHU	III	CLS, NPR, PRI	P
	91.7	WXCI	I	DVRS	C
AM	800	WLAD	III	NWS, TLK	

HARTFORD, CT

	FREQ	STATION	POWER	FORMAT	
	96.5	WTIC	IIIIIIII	AC	
	100.5	WRCH	IIIIIIII	LGHT AC	
	95.7	WKSS	IIIIIIII	TOP 40	
	93.7	WZMX	IIIIIIII	70's RK	
	105.9	WHCN	IIIIIIII	CLS RK	
	106.9	WCCC	IIIIIIII	ROCK	
	102.9	WDRC	III	OLDS	
AM	840	WRYM	III	SPAN	
AM	1120	WPRX	III	SPAN	
AM	1230	WLAT	I	SPAN	
AM	1470	WMMW	III	SPAN	
AM	1360	WDRC	IIIII	BB, NOS	
	88.1	WESU	I	DVRS	P
	89.3	WRTC	I	DVRS	C
	90.5	WPKT	III	CLS, NPR, PRI	P
	91.3	WWUH	III	DVRS	C
	91.7	WHUS	III	DVRS	C
AM	1550	WDZK	IIIII	KIDS	
AM	910	WNEZ	IIIIIIII	URBAN	
AM	1080	WTIC	IIIIIIII	NWS, TLK	
AM	1410	WPOP	IIIII	SPRTS	

NEW HAVEN, CT

	FREQ	STATION	POWER	FORMAT	
	101.3	WKCI	IIIII	TOP 40	
	99.1	WPLR	III	CLS RK	
AM	1220	WQUN	I	BB, NOS	
AM	1300	WAVZ	I	BB, NOS	
	88.7	WNHU	I	DVRS	C
	94.3	WYBC	I	URBAN	
AM	960	WELI	IIIIIIII	NWS, TLK	

NEW LONDON, CT

	FREQ	STATION	POWER	FORMAT	
	100.9	WTYD	III	AC	
	106.5	WBMW	III	LGHT AC	
AM	1310	WICH	IIIII	LGHT AC	
	105.5	WQGN	III	TOP 40	
	102.3	WVVE	III	OLDS	
	97.7	WCTY	III	CTRY	
	98.7	WNLC	III	BB, NOS	
	107.7	WKCD	I	JZZ	
	91.1	WCNI	I	DVRS	C

STAMFORD, CT

	FREQ	STATION	POWER	FORMAT	
	107.9	WEBE	IIIII	AC	
AM	1490	WGCH	I	AC	
	95.9	WEFX	III	CLS RK	

	FREQ	STATION	POWER	FORMAT	
	96.7	WKHL	III	OLDS	
	88.5	WEDW	I	CLS, NPR, PRI	P
AM	1350	WNLK	I	NWS, TLK	
AM	1400	WSTC	IIIII	NWS, TLK	

WATERBURY, CT

	FREQ	STATION	POWER	FORMAT	
	104.1	WMRQ	IIIIIIII	MOD RK	
	92.5	WWYZ	IIIIIIII	CTRY	
AM	1240	WWCO	I	BB, NOS	
AM	1320	WATR	IIIII	NWS, TLK	

BOSTON, MA

	FREQ	STATION	POWER	FORMAT	
	98.5	WBMX	IIIIIIII	AC	
	106.7	WMJX	IIIIIIII	AC	
	107.9	WXKS	IIIIIIII	TOP 40	
	92.5	WXRV	IIIIIIII	ALT RK	
	92.9	WBOS	IIIIIIII	ALT RK	
	93.7	WEGQ	IIIII	70's RK	
	100.7	WZLX	III	CLS RK	
	101.7	WFNX	III	DVRS	
	104.1	WBCN	IIIIIIII	MOD RK	
	103.3	WODS	III	OLDS	
	105.7	WROR	IIIIIIII	OLDS	
	97.7	WCAV	III	CTRY	
	99.5	WKLB	IIIIIIII	CTRY	
AM	1330	WRCA	IIIII	SPAN	
	96.9	WSJZ	IIIIIIII	JZZ	
	99.1	WPLM	IIIII	JZZ	
	102.5	WCRB	IIIIIIII	CLS	
	104.9	WBOQ	III	CLS	
AM	1230	WESX	I	BB, NOS	
AM	1390	WPLM	IIIII	JZZ	
AM	1400	WLLH	III	BB, NOS	
AM	1430	WXKS	IIIII	BB, NOS	
	88.1	WMBR	III	DVRS	P
	88.3	WGAO	I	AC	C
	88.9	WERS	III	DVRS	C
	89.7	WGBH	IIIIIIII	CLS, NPR, PRI	P
	90.3	WZBC	I	ROCK	C
	90.9	WBUR	III	NWS, TLK, NPR, PRI	P
	91.3	WDJM	I	DVRS	C
	91.5	WJUL	I	DVRS	C
	91.5	WBIM	I	DVRS	C
	91.5	WMFO	I	DVRS	C
	91.5	WMLN	I	DVRS	C
	91.7	WMWM	I	DVRS	C
AM	950	WROL	IIIIIIII	RLG AC	
	91.9	WUMB	I	ETH, NPR	P
AM	590	WEZE	IIIIIIII	RLG	
AM	1260	WPZE	IIIII	KIDS	
AM	1460	WBET	IIIII	FS	
	94.5	WJMN	IIIIIIII	URBAN	
AM	680	WRKO	IIIIIIII	NWS, TLK	
AM	800	WCCM	III	NWS, TLK	
AM	850	WEEI	IIIIIIII	SPRTS	
AM	980	WCAP	IIIIIIII	NWS, TLK	
AM	1030	WBZ	IIIIIIII	NWS, TLK	
AM	1510	WNRB	IIIIIIII	SPRTS	

CAPE COD, MA

	FREQ	STATION	POWER	FORMAT	
	99.9	WQRC	IIIII	LGHT AC	
	101.9	WCIB	IIIII	AC	

3

MA
ME

	FREQ	STATION	POWER	FORMAT	
	106.1	WCOD	IIIII	AC	
	93.5	WYST	III	TOP 40	
	101.1	WWKJ	III	CLS RK	
	102.9	WPXC	IIIII	MOD RK	
	104.7	WKPE	IIIIIIII	MOD RK	
	103.9	WOCN	III	BB, NOS	
	107.5	WFCC	IIIII	CLS	
	91.9	WFPB	III	JZZ, NPR	C
	92.1	WOMR	I	DVRS	P
AM	1240	WBUR	I	NWS, TLK, NPR, PRI	C
	95.1	WXTK	IIIII	NWS, TLK	

SPRINGFIELD, MA

	FREQ	STATION	POWER	FORMAT	
	93.1	WHYN	III	AC	
	94.7	WMAS	III	AC	
	99.3	WHMP	III	MOD RK	
	100.9	WRNX	III	ALT RK	
	102.1	WAQY	IIIIIIII	CLS RK	
AM	1270	WSPR	IIIII	SPAN	
AM	1490	WACM	I	SPAN	
AM	1250	WARE	IIIII	BB, NOS	
AM	1450	WMAS	I	BB, NOS	
	88.5	WFCR	III	DVRS, NPR, PRI	P
	89.3	WAMH	III	DVRS	C
	89.5	WSKB	I	DVRS	C
	90.7	WTCC	I	DVRS	C
	91.1	WMUA	I	DVRS	C
	91.5	WMHC	I	DVRS	C
	91.9	WOZQ	I	DVRS	C
	91.9	WAIC	I	URBAN	C
AM	730	WACE	IIIIIIII	RLG	
AM	560	WHYN	IIIIIIII	NWS, TLK	
AM	640	WNNZ	IIIIIIII	NWS, TLK	
AM	1400	WHMP	I	NWS, TLK	

WORCESTER, MA

	FREQ	STATION	POWER	FORMAT	
	96.1	WSRS	III	AC	
	104.5	WXLO	IIIIIIII	HOT AC	
	98.9	WORC	III	CLS RK	
	100.1	WQVR	III	CTRY	
	88.1	WCHC	I	AC	C
	90.5	WICN	III	JZZ, NPR, PRI	P
	91.3	WCUW	I	DVRS	P
	91.9	WBPR	III	ETH, NPR	P
AM	580	WTAG	IIIIIIII	NWS, TLK	
AM	1310	WORC	IIIII	NWS, TLK	
AM	1440	WWTM	IIIII	SPRTS	

AUGUSTA. ME

	FREQ	STATION	POWER	FORMAT	
	92.3	WMME	IIIII	TOP 40	
	104.3	WABK	IIIII	OLDS	
	98.5	WEBB	IIIII	CTRY	
	101.3	WKCG	IIIII	CTRY	
AM	1280	WFAU	IIIII	BB, NOS	
	90.5	WMHB	I	DVRS	P
	91.3	WMEW	III	CLS, NPR, PRI	P

BANGOR, ME

	FREQ	STATION	POWER	FORMAT	
	92.9	WEZQ	III	AC	
	97.1	WWBX	IIIIIIII	TOP 40	
	107.3	WBZN	IIIII	TOP 40	
	100.3	WKIT	IIIIIIII	ROCK	

	FREQ	STATION	POWER	FORMAT	
	106.5	WQCB	‖‖‖‖‖	CTRY	
AM	910	WABI	‖‖‖‖‖	BB, NOS	
	89.3	WHSN	I	AC	C
	90.9	WMEH	III	CLS, NPR, PRI	P
	103.9	WVOM	‖‖‖‖‖	NWS, TLK	
AM	620	WZON	‖‖‖‖‖	SPRTS	

PORTLAND, ME

	FREQ	STATION	POWER	FORMAT	
	99.9	WMWX	‖‖‖	AC	
	97.9	WJBQ	‖‖‖‖‖	TOP 40	
	93.1	WMGX	‖‖‖	CLS RK	
	93.9	WCYI	‖‖‖‖‖	DVRS	
	98.9	WCLZ	‖‖‖	ALT RK	
	102.9	WBLM	‖‖‖‖‖	ROCK	
	100.9	WYNZ	III	OLDS	
	101.9	WPOR	‖‖‖	CTRY	
	107.5	WTHT	‖‖‖‖‖	CTRY	
	106.3	WBQW	III	CLS	
	106.7	WLAM	III	BB, NOS	
AM	870	WLAM	‖‖‖‖‖	BB, NOS	
	90.9	WMPG	I	DVRS	C
	91.1	WBOR	I	DVRS	C
	91.5	WRBC	I	DVRS	C
	91.5	WSJB	I	TOP 40	C
AM	560	WGAN	‖‖‖‖‖	NWS, TLK	
AM	970	WZAN	‖‖‖‖‖	NWS, TLK	
AM	1440	WJAE	‖‖‖	SPRTS	

MANCHESTER, NH

	FREQ	STATION	POWER	FORMAT	
	95.7	WZID	III	AC	
	101.1	WGIR	‖‖‖	ROCK	
AM	1370	WFEA	‖‖‖	BB, NOS	
AM	1320	WDER	‖‖‖	RLG AC	
AM	610	WGIR	‖‖‖‖‖	NWS, TLK	
AM	1250	WKBR	‖‖‖	NWS, TLK	

PORTSMOUTH, NH

	FREQ	STATION	POWER	FORMAT	
	95.3	WXHT	III	AC	
	98.7	WBYY	III	LGHT AC	
	107.1	WERZ	III	TOP 40	
	92.1	WCDQ	III	CLS RK, ALT RK	
	100.3	WHEB	‖‖‖	CLS RK	
	102.1	WXBP	III	CLS RK	
	105.3	WXBB	III	CLS RK	
	96.7	WQSO	III	OLDS	
	97.5	WOKQ	‖‖‖	CTRY	
AM	930	WZNN	‖‖‖‖‖	BB, NOS	
AM	1540	WMYF	‖‖‖	BB, NOS	
	91.3	WUNH	III	DVRS	C
AM	1340	WWNH	I	RLG AC	
AM	1270	WTSN	‖‖‖	NWS, TLK	

ALBANY, NY

FREQ	STATION	POWER	FORMAT
95.5	WYJB	‖‖‖‖‖	AC
100.9	WKLI	III	HOT AC
101.3	WJKE	III	AC
104.5	WXLE	III	HOT AC
92.3	WFLY	‖‖‖‖‖	TOP 40
99.5	WRVE	III	CLS RK
102.3	WXCR	III	CLS RK
103.5	WQBJ	III	DVRS
103.9	WQBK	III	DVRS

ME
NH
NY

5

	FREQ	STATION	POWER	FORMAT	
	106.5	WPYX	IIIIIIII	ROCK	
	98.3	WTRY	III	OLDS	
AM	980	WTRY	IIIIIIII	OLDS	
	93.5	WZZM	III	CTRY	
	96.3	WPTR	III	CTRY	
	101.1	WBUG	III	CTRY	
	107.7	WGNA	III	CTRY	
AM	1460	WGNA	IIIIIIII	CTRY	
AM	1570	WBUG	I	CTRY	
	94.5	WABY	III	BB, NOS	
	103.1	WHRL	III	JZZ	
AM	1400	WABY	I	BB, NOS	
	88.3	WVCR	III	DVRS	C
	89.1	WMHT	III	CLS, NPR, PRI	P
	89.7	WRUC	I	ROCK	C
	90.3	WAMC	IIIII	NWS, TLK, NPR, PRI	P
	90.9	WCDB	I	ROCK	C
	91.1	WSPN	I	DVRS	C
	91.5	WRPI	III	DVRS	C
	96.7	WDCD	III	RLG AC	
AM	1540	WDCD	IIIIIIII	RLG AC	
AM	1490	WCSS	I	FS	
AM	590	WROW	IIIIIIII	NWS, TLK	
AM	810	WGY	IIIIIIII	NWS, TLK	
AM	1300	WTMM	IIIII	SPRTS	

BINGHAMTON, NY

	FREQ	STATION	POWER	FORMAT	
	100.5	WCDW	III	AC	
	101.7	WLTB	III	LGHT AC	
	102.1	WAVR	III	AC	
	105.7	WMRV	IIIII	TOP 40	
	99.1	WAAL	IIIIIIII	CLS RK	
	103.3	WMXW	III	OLDS	
	104.1	WYOS	I	OLDS	
	98.1	WHWK	III	CTRY	
AM	680	WINR	IIIII	BB, NOS	
AM	1360	WKOP	IIIII	BB, NOS	
	89.3	WSKG	III	CLS, NPR, PRI	P
	90.5	WHRW	I	DVRS	C
	91.5	WSQX	I	NWS, TLK, NPR	P
AM	1290	WNBF	IIIII	NWS, TLK	
AM	1430	WENE	IIIII	SPRTS	

BUFFALO, NY

	FREQ	STATION	POWER	FORMAT	
	92.9	WLCE	IIIIIIII	HOT AC	
	96.1	WJYE	IIIII	LGHT AC	
	102.5	WMJQ	IIIIIIII	HOT AC	
	98.5	WKSE	III	TOP 40	
	96.9	WGRF	III	CLS RK	
	103.3	WEDG	IIIII	DVRS	
	104.1	WHTT	IIIIIIII	OLDS	
	106.5	WYRK	IIIII	CTRY	
AM	1300	WXRL	III	CTRY	
AM	1230	WECK	I	BB, NOS	
	88.7	WBFO	IIIII	NWS, TLK, NPR, PRI	P
	89.7	WNJA	III	CLS, NPR, PRI	P
	91.3	WBNY	I	DVRS	C
	94.5	WNED	IIIIIIII	CLS, NPR, PRI	P
AM	970	WNED	IIIIIIII	NWS, NPR, PRI	P
	99.5	WDCX	IIIIIIII	RLG AC	
AM	1080	WUFO	I	GOSP	
AM	1270	WHLD	IIIII	RLG	

6

	FREQ	STATION	POWER	FORMAT	
	93.7	WBLK	IIIII	URBAN	
AM	1400	WWWS	I	URBAN	
AM	550	WGR	IIIIIIII	NWS, TLK	
AM	930	WBEN	IIIIIIII	NWS, TLK	
AM	1340	WLVL	I	NWS, TLK	
AM	1520	WWKB	IIIIIIII	SPRTS	

ELMIRA, NY

	FREQ	STATION	POWER	FORMAT	
	92.7	WENY	III	AC	
	98.3	WVIN	III	AC	
	98.7	WCBA	III	AC	
	106.1	WNKI	IIIIIIII	AC	
	94.3	WLVY	III	TOP 40	
	105.3	WKPQ	IIIIIIII	DVRS	
	97.7	WGMM	III	OLDS	
AM	1380	WABH	III	OLDS	
AM	1590	WEHH	I	OLDS	
	92.1	WCKR	III	CTRY	
	99.5	WOKN	III	CTRY	
	100.9	WPGI	I	CTRY	
AM	1350	WCBA	I	BB, NOS	
AM	820	WWLZ	IIIIIIII	NWS, TLK	
AM	1230	WENY	I	NWS, TLK	
AM	1320	WHHO	IIIII	NWS, TLK	
AM	1410	WELM	IIIII	SPRTS	
AM	1450	WCLI	I	NWS, TLK	

ITHACA, NY

	FREQ	STATION	POWER	FORMAT	
	97.3	WYXL	IIIII	AC	
	93.5	WVBR	I	ROCK	C
AM	1470	WTKO	IIIII	OLDS	
	103.7	WQNY	III	CTRY	
	90.9	WSQG	III	CLS, NPR, PRI	P
	91.7	WICB	I	DVRS	C
AM	870	WHCU	IIIIIIII	NWS, TLK	

LONG ISLAND, NY

	FREQ	STATION	POWER	FORMAT	
	97.5	WALK	IIIII	AC	
	98.3	WKJY	III	AC	
	101.7	WBAZ	III	AC	
AM	1370	WALK	I	LGHT AC	
	103.5	WKTU	III	TOP 40	
	106.1	WBLI	IIIII	TOP 40	
	95.3	WHFM	III	ROCK	
	96.7	WEHM	III	DVRS	
	98.5	WDRE	III	DVRS	
	102.3	WBAB	III	CLS RK	
	103.9	WRCN	III	ROCK	
	92.1	WLNG	III	OLDS	
	103.1	WBZO	III	OLDS	
	94.3	WMJC	III	CTRY	
AM	540	WLUX	IIIII	BB, NOS	
	88.3	WPBX	III	DVRS, NPR, PRI	P
	88.7	WRHU	I	DVRS	C
	89.9	WSUF	III	CLS, NPR, PRI	P
	90.1	WUSB	I	DVRS	C
	90.3	WHCP	I	LGHT AC	C

NEW YORK CITY, NY

	FREQ	STATION	POWER	FORMAT	
	98.3	WMGQ	III	AC	
	98.7	WRKS	IIIIIIII	AC	
	100.7	WHUD	IIIII	AC	
	103.9	WFAS	III	AC	

	FREQ	STATION	POWER	FORMAT	
	105.1	WBIX	IIIIIIII	HOT AC	
	106.7	WLTW	IIIIIIII	LGHT AC	
	95.5	WPLJ	III	TOP 40	
	97.1	WQHT	III	TOP 40	
	100.3	WHTZ	III	TOP 40	
	92.3	WXRK	III	MOD RK	
	102.7	WNEW	III	CLS RK	
	104.3	WAXQ	IIIIIIII	CLS RK	
	101.1	WCBS	III	OLDS	
	107.1	WWXY	III	CTRY	
	93.1	WPAT	III	SPAN	
	97.9	WSKQ	III	SPAN	
	105.9	WCAA	III	SPAN	
AM	1280	WADO	IIIII	SPAN	
AM	1600	WWRL	IIIII	BLK, R&B	
	96.3	WQXR	III	CLS	
	101.9	WQCD	III	JZZ	
AM	1560	WQEW	IIIIIIII	BB, NOS	
	88.3	WBGO	III	JZZ, NPR	P
	88.7	WRSU	I	DVRS	C
	89.1	WFDU	III	DVRS	C
	89.1	WNYU	III	DVRS	C
	89.5	WSOU	III	DVRS	C
	89.9	WKCR	III	DVRS	C
	90.3	WRPR	I	DVRS	C
	90.7	WFUV	III	DVRS, NPR, PRI	C
	91.1	WFMU	III	DVRS	P
	93.9	WNYC	III	CLS, NPR, PRI	P
	99.5	WBAI	IIIIIIII	DVRS	P
AM	820	WNYC	IIIIIIII	NWS, TLK, NPR, PRI	P
AM	970	WWDJ	IIIIIIII	RLG AC	
AM	570	WMCA	IIIIIIII	RLG	
	107.5	WBLS	IIIIIIII	URBAN	
AM	930	WPAT	IIIIIIII	NWS, TLK	
AM	620	WJWR	IIIIIIII	SPRTS	
AM	660	WFAN	IIIIIIII	SPRTS	
AM	710	WOR	IIIIIIII	NWS, TLK	
AM	770	WABC	IIIIIIII	NWS, TLK	
AM	880	WCBS	IIIIIIII	NWS	
AM	1010	WINS	IIIIIIII	NWS	
AM	1050	WEVD	IIIIIIII	NWS, TLK	
AM	1130	WBBR	IIIIIIII	NWS	

NEWBURGH, NY

	FREQ	STATION	POWER	FORMAT
	96.7	WTSX	III	AC
	103.1	WGNY	III	HOT AC
	92.7	WRRV	III	DVRS
AM	1220	WGNY	I	OLDS
AM	1340	WALL	I	NWS, TLK

POUGHKEEPSIE, NY

	FREQ	STATION	POWER	FORMAT	
	92.1	WRNQ	III	AC	
	104.7	WSPK	IIIIIIII	TOP 40	
	96.9	WDSP	III	DVRS	
	101.5	WPDH	III	ROCK	
	97.7	WCZX	III	OLDS	
	96.1	WTND	I	CTRY	
AM	1390	WEOK	IIIII	BB, NOS	
	88.7	WRHV	I	CLS, NPR, PRI	P
	91.3	WVKR	I	DVRS	C
AM	1260	WBNR	I	NWS, TLK	
AM	1450	WKIP	I	NWS, TLK	

8

	FREQ	STATION	POWER	FORMAT	
	93.9	WDNY	III	AC	
	100.5	WVOR	IIIII	HOT AC	
	101.3	WRMM	IIIII	AC	
	102.3	WISY	III	LGHT AC	
	106.7	WYSY	III	LGHT AC	
	97.9	WPXY	IIIII	TOP 40	
	107.3	WMAX	I	TOP 40	
	93.3	WQRV	III	CLS RK	
	93.7	WLLW	III	CLS RK	
	95.1	WNVE	IIIIIIII	MOD RK	
	96.5	WCMF	IIIII	ROCK	
	103.5	WNNR	I	CLS RK	
	98.9	WBBF	IIIII	OLDS	
	92.5	WBEE	IIIII	CTRY	
	105.9	WJZR	I	JZZ	
AM	950	WEZO	III	BB, NOS	
	88.5	WRUR	I	DVRS	C
	89.1	WBSU	I	AC	C
	89.7	WEOS	III	DVRS	C
	89.7	WITR	I	DVRS	C
	90.1	WGMC	I	JZZ	P
	90.7	WGCC	I	DVRS	C
	91.5	WXXI	III	CLS, NPR, PRI	P
AM	1370	WXXI	IIIII	NWS, TLK, NPR, PRI	P
AM	1460	WWWG	IIIII	GOSP	
	89.3	WGSU	I	EDU	C
	99.7	WZXV	III	RLG	
	102.7	WDCZ	III	RLG	
	103.9	WDKX	III	URBAN	
AM	1040	WYSL	III	NWS	
AM	1180	WHAM	IIIIIIII	NWS, TLK	
AM	1280	WHTK	IIIII	NWS, TLK	
AM	1490	WBTA	I	NWS	

SYRACUSE, NY

	FREQ	STATION	POWER	FORMAT	
	94.5	WYYY	IIIIIIII	HOT AC	
	102.1	WRDS	I	AC	
	105.9	WLTI	III	LGHT AC	
	106.3	WMCR	III	AC	
	107.9	WWHT	IIIII	HOT AC	
	93.1	WNTQ	IIIIIIII	TOP 40	
	95.7	WAQX	III	ROCK	
	99.5	WTKW	III	CLS RK	
	100.9	WKRL	I	MOD RK	
	105.5	WTKV	I	CLS RK	
	92.1	WSEN	III	OLDS	
	101.7	WSCP	III	CTRY	
	104.7	WBBS	IIIII	CTRY	
AM	1200	WTLA	I	BB, NOS	
	88.3	WAER	III	JZZ, NPR, PRI	P
	88.9	WITC	I	DVRS	C
	88.9	WNYO	I	DVRS	C
	89.1	WJPZ	I	TOP 40	P
	89.9	WRVO	III	NWS, TLK, NPR, PRI	P
	90.1	WRCU	I	DVRS	C
	91.3	WCNY	III	CLS, NPR, PRI	P
AM	1390	WDCW	IIIII	RLG AC	
AM	1540	WSIV	I	RLG	
AM	570	WSYR	IIIIIIII	NWS, TLK	
AM	620	WHEN	IIIIIIII	SPRTS	
AM	1260	WNSS	IIIII	NWS	
AM	1490	WOLF	I	NWS, TLK	

NY

9

NY
RI

	FREQ	STATION	POWER	FORMAT	
	93.5	WRFM	I	AC	
	98.7	WLZW	IIIII	AC	
	97.9	WOWZ	III	TOP 40	
	102.5	WSKS	IIIIIIII	TOP 40	
	105.5	WOWB	I	TOP 40	
	94.9	WKLL	IIIII	MOD RK	
	96.9	WOUR	III	ROCK	
	107.3	WRCK	IIIII	CLS RK	
	92.7	WXUR	III	OLDS	
	96.1	WODZ	III	OLDS	
AM	1450	WODZ	I	OLDS	
	101.3	WBRV	III	CTRY	
	104.3	WFRG	IIIIIIII	CTRY	
AM	1150	WRUN	IIIII	CTRY	
AM	1230	WLFH	I	CTRY	
AM	1310	WTLB	IIIII	BB, NOS	
AM	1480	WADR	IIIII	BB, NOS	
	89.5	WUNY	III	CLS, NPR, PRI	P
	90.7	WPNR	I	DVRS	C
	91.5	WVHC	I	DVRS	C
	100.7	WVVC	I	RLG AC	
AM	1550	WUTQ	I	EZ	
AM	950	WIBX	IIIIIIII	NWS, TLK	

WATERTOWN, NY

	FREQ	STATION	POWER	FORMAT	
	93.3	WCIZ	III	AC	
	103.1	WTOJ	III	AC	
AM	790	WTNY	IIIII	AC	
	106.7	WWLF	IIIIIIII	TOP 40	
	100.7	WOTT	III	OLDS	
AM	1410	WUZZ	IIIII	OLDS	
	97.5	WFRY	IIIII	CTRY	
	90.9	WJNY	III	CLS, NPR, PRI	P
AM	1240	WATN	I	NWS, TLK	

PROVIDENCE, RI

	FREQ	STATION	POWER	FORMAT	
	93.3	WSNE	IIIII	HOT AC	
	105.1	WWLI	IIIII	AC	
AM	550	WLKW	IIIII	LGHT AC	
	92.3	WPRO	IIIII	TOP 40	
	102.7	WAKX	I	TOP 40	
	106.3	WWKX	III	TOP 40	
	107.1	WFHN	III	TOP 40	
	94.1	WHJY	IIIII	ROCK	
	95.5	WBRU	III	MOD RK	
	99.7	WXEX	III	DVRS	
	100.3	WHKK	III	OLDS	
	103.7	WWRX	IIIIIIII	CLS RK	
	101.5	WWBB	III	OLDS	
AM	1240	WOON	I	OLDS	
	98.1	WCTK	IIIIIIII	CTRY	
AM	1340	WNBH	I	CTRY	
AM	1320	WJYT	IIIII	SPAN	
	88.1	WBLQ	I	CLS RK	P
	88.3	WQRI	I	DVRS	C
	88.7	WJMF	I	DVRS	C
	90.3	WRIU	III	DVRS	C
	91.1	WSMU	III	JZZ	C
	91.3	WDOM	I	AC	C
	91.3	WSHL	I	DVRS	C
	97.3	WJFD	IIIII	ETH	
AM	1400	WHTB	I	ETH	

10

	FREQ	STATION	POWER	FORMAT	
AM	1450	WDYZ	I	KIDS	
AM	1590	WARV	IIIII	RLG	
AM	630	WPRO	IIIIIIII	NWS, TLK	
AM	920	WHJJ	IIIIIIII	NWS, TLK	
AM	990	WALE	IIIIIIII	NWS, TLK	
AM	1230	WERI	I	NWS, TLK	
AM	1290	WRNI	IIIII	NWS	
AM	1420	WBSM	IIIII	NWS, TLK	
AM	1480	WSAR	IIIII	NWS, TLK	
AM	1570	WPEP	I	NWS, TLK	

BURLINGTON, VT

	FREQ	STATION	POWER	FORMAT	
	92.9	WEZF	IIIIIIII	AC	
	95.5	WXXX	III	TOP 40	
	98.9	WOKO	IIIII	CTRY	
AM	1230	WJOY	I	BB, NOS	
	90.1	WRUV	I	DVRS	C
	107.9	WVPS	IIIIIIII	CLS, NPR, PRI	P
AM	620	WVMT	IIIIIIII	NWS, TLK	
AM	1390	WKDR	IIIII	NWS, TLK	

RI
VT

11

88.1
CT	HARTFORD	WESU	2
MA	BOSTON	WMBR	3
MA	WORCESTER	WCHC	4
RI	PROVIDENCE	WBLQ	10

88.3
MA	BOSTON	WGAO	3
NY	ALBANY	WVCR	6
NY	LONG ISLAND	WPBX	7
NY	NEW YORK CITY	WBGO	8
NY	SYRACUSE	WAER	9
RI	PROVIDENCE	WQRI	11

88.5
CT	DANBURY	WVOF	2
CT	STAMFORD	WEDW	3
MA	SPRINGFIELD	WFCR	4
NY	ROCHESTER	WRUR	9

88.7
CT	NEW HAVEN	WNHU	2
NY	BUFFALO	WBFO	6
NY	LONG ISLAND	WRHU	7
NY	NEW YORK CITY	WRHV	8
NY	POUGHKEEPSIE	WRHV	9
RI	PROVIDENCE	WJMF	11

88.9
MA	BOSTON	WERS	3
NY	SYRACUSE	WITC	9
NY	SYRACUSE	WNYO	9

89.1
NY	ALBANY	WMHT	6
NY	NEW YORK CITY	WFDU	8
NY	NEW YORK CITY	WNYU	8
NY	ROCHESTER	WBSU	9
NY	SYRACUSE	WJPZ	9

89.3
CT	HARTFORD	WRTC	2
MA	SPRINGFIELD	WAMH	4
ME	BANGOR	WHSN	5
NY	BINGHAMTON	WSKG	6
NY	ROCHESTER	WGSU	9

89.5
CT	BRIDGEPORT	WPKN	2
MA	SPRINGFIELD	WSKB	4
NY	NEW YORK CITY	WSOU	8
NY	UTICA	WUNY	10

89.7
MA	BOSTON	WGBH	3
NY	ALBANY	WRUC	6
NY	BUFFALO	WNJA	6
NY	ROCHESTER	WEOS	9
NY	ROCHESTER	WITR	9

89.9
NY	LONG ISLAND	WSUF	7
NY	NEW YORK CITY	WKCR	8
NY	SYRACUSE	WRVO	9

90.1
NY	LONG ISLAND	WUSB	7
NY	ROCHESTER	WGMC	9
NY	SYRACUSE	WRCU	9
VT	BURLINGTON	WRUV	11

90.3
MA	BOSTON	WZBC	3
NY	ALBANY	WAMC	6
NY	LONG ISLAND	WHCP	7
NY	NEW YORK CITY	WRPR	8
RI	PROVIDENCE	WRIU	11

90.5
CT	HARTFORD	WPKT	2
MA	WORCESTER	WICN	4
ME	AUGUSTA	WMHB	4
NY	BINGHAMTON	WHRW	6

90.7
MA	SPRINGFIELD	WTCC	4
NY	NEW YORK CITY	WFUV	8
NY	ROCHESTER	WGCC	9
NY	UTICA	WPNR	10

90.9
MA	BOSTON	WBUR	3
ME	BANGOR	WMEH	5
ME	PORTLAND	WMPG	5
NY	ALBANY	WCDB	6
NY	ITHACA	WSQG	7
NY	WATERTOWN	WJNY	10

91.1
CT	DANBURY	WSHU	2
CT	NEW LONDON	WCNI	2
MA	SPRINGFIELD	WMUA	4
ME	PORTLAND	WBOR	5
NY	ALBANY	WSPN	6
NY	NEW YORK CITY	WFMU	8
RI	PROVIDENCE	WSMU	11

91.3
CT	HARTFORD	WWUH	2
MA	BOSTON	WDJM	3
MA	WORCESTER	WCUW	4
ME	AUGUSTA	WMEW	4
NH	PORTSMOUTH	WUNH	5
NY	BUFFALO	WBNY	6
NY	POUGHKEEPSIE	WVKR	9
NY	SYRACUSE	WCNY	9
RI	PROVIDENCE	WDOM	11
RI	PROVIDENCE	WSHL	11

91.5
MA	BOSTON	WJUL	3
MA	BOSTON	WBIM	3
MA	BOSTON	WMFO	3
MA	BOSTON	WMLN	3
MA	SPRINGFIELD	WMHC	4
ME	PORTLAND	WRBC	5
ME	PORTLAND	WSJB	5
NY	ALBANY	WRPI	6
NY	BINGHAMTON	WSQX	6
NY	ROCHESTER	WXXI	9
NY	UTICA	WVHC	10

91.7
CT	DANBURY	WXCI	2
CT	HARTFORD	WHUS	2
MA	BOSTON	WMWM	3
NY	ITHACA	WICB	7

91.9
MA	BOSTON	WUMB	3
MA	CAPE COD	WFPB	4
MA	SPRINGFIELD	WOZQ	4
MA	SPRINGFIELD	WAIC	4
MA	WORCESTER	WBPR	4

92.1
MA	CAPE COD	WOMR	4
NH	PORTSMOUTH	WCDQ	5
NY	ELMIRA	WCKR	7
NY	LONG ISLAND	WLNG	7
NY	POUGHKEEPSIE	WRNQ	8
NY	SYRACUSE	WSEN	9

92.3
ME	AUGUSTA	WMME	4
NY	ALBANY	WFLY	6
NY	NEW YORK CITY	WXRK	8
RI	PROVIDENCE	WPRO	10

92.5
CT	WATERBURY	WWYZ	3
MA	BOSTON	WXRV	3
NY	ROCHESTER	WBEE	9

92.7
NY	ELMIRA	WENY	7
NY	NEWBURGH	WRRV	8
NY	UTICA	WXUR	10

92.9
MA	BOSTON	WBOS	3
ME	BANGOR	WEZQ	5
NY	BUFFALO	WLCE	6
VT	BURLINGTON	WEZF	11

93.1
MA	SPRINGFIELD	WHYN	4
ME	PORTLAND	WMGX	5
NY	NEW YORK CITY	WPAT	8
NY	SYRACUSE	WNTQ	9

93.3
NY	ROCHESTER	WQRV	9
NY	WATERTOWN	WCIZ	10
RI	PROVIDENCE	WSNE	10

93.5
MA	CAPE COD	WYST	4
NY	ALBANY	WZZM	6
NY	ITHACA	WVBR	7
NY	UTICA	WRFM	10

93.7
CT	HARTFORD	WZMX	2

93.9 (continued from previous)

MA	BOSTON	WEGQ	3
NY	BUFFALO	WBLK	7
NY	ROCHESTER	WLLW	9

93.9
ME	PORTLAND	WCYI	5
NY	NEW YORK CITY	WNYC	8
NY	ROCHESTER	WDNY	9

94.1
RI	PROVIDENCE	WHJY	10

94.3
CT	NEW HAVEN	WYBC	2
NY	ELMIRA	WLVY	7
NY	LONG ISLAND	WMJC	7

94.5
MA	BOSTON	WJMN	3
NY	ALBANY	WABY	6
NY	BUFFALO	WNED	6
NY	SYRACUSE	WYYY	9

94.7
MA	SPRINGFIELD	WMAS	4

94.9
NY	UTICA	WKLL	10

95.1
CT	DANBURY	WRKI	2
MA	CAPE COD	WXTK	4
NY	ROCHESTER	WNVE	9

95.3
NH	PORTSMOUTH	WXHT	5
NY	LONG ISLAND	WHFM	7

95.5
NY	ALBANY	WYJB	5
NY	NEW YORK CITY	WPLJ	8
RI	PROVIDENCE	WBRU	10
VT	BURLINGTON	WXXX	11

95.7
CT	HARTFORD	WKSS	2
NH	MANCHESTER	WZID	5
NY	SYRACUSE	WAQX	9

95.9
CT	STAMFORD	WEFX	2

96.1
MA	WORCESTER	WSRS	4
NY	BUFFALO	WJYE	6
NY	POUGHKEEPSIE	WTND	8
NY	UTICA	WODZ	10

96.3
NY	ALBANY	WPTR	6
NY	NEW YORK CITY	WQXR	8

96.5
CT	HARTFORD	WTIC	2
NY	ROCHESTER	WCMF	9

96.7
CT	STAMFORD	WKHL	3
NH	PORTSMOUTH	WQSO	5
NY	ALBANY	WDCD	6
NY	LONG ISLAND	WEHM	7
NY	NEWBURGH	WTSX	8

96.9
MA	BOSTON	WSJZ	3
NY	BUFFALO	WGRF	6
NY	POUGHKEEPSIE	WDSP	8
NY	UTICA	WOUR	10

97.1
ME	BANGOR	WWBX	5
NY	NEW YORK CITY	WQHT	8

97.3
NY	ITHACA	WYXL	7
RI	PROVIDENCE	WJFD	11

97.5
NH	PORTSMOUTH	WOKQ	5
NY	LONG ISLAND	WALK	7
NY	WATERTOWN	WFRY	10

97.7
CT	NEW LONDON	WCTY	2
MA	BOSTON	WCAV	3
NY	ELMIRA	WGMM	7
NY	POUGHKEEPSIE	WCZX	8

97.9
ME	PORTLAND	WJBQ	5
NY	NEW YORK CITY	WSKQ	8
NY	ROCHESTER	WPXY	9
NY	UTICA	WOWZ	10

98.1
NY	BINGHAMTON	WHWK	6
RI	PROVIDENCE	WCTK	10

98.3
CT	DANBURY	WDAQ	2
NY	ALBANY	WTRY	6
NY	ELMIRA	WVIN	7
NY	LONG ISLAND	WKJY	7
NY	NEW YORK CITY	WMGQ	8

98.5
MA	BOSTON	WBMX	3
ME	AUGUSTA	WEBB	4
NY	BUFFALO	WKSE	6
NY	LONG ISLAND	WDRE	7

98.7
CT	NEW LONDON	WNLC	2
NH	PORTSMOUTH	WBYY	5
NY	ELMIRA	WCBA	7
NY	NEW YORK CITY	WRKS	8
NY	UTICA	WLZW	10

98.9
MA	WORCESTER	WORC	4
ME	PORTLAND	WCLZ	5
NY	ROCHESTER	WBBF	9
VT	BURLINGTON	WOKO	11

99.1
CT	NEW HAVEN	WPLR	2
MA	BOSTON	WPLM	3
NY	BINGHAMTON	WAAL	6

99.3
MA	SPRINGFIELD	WHMP	4

99.5
MA	BOSTON	WKLB	3
NY	ALBANY	WRVE	6
NY	BUFFALO	WDCX	7
NY	ELMIRA	WOKN	7
NY	NEW YORK CITY	WBAI	8
NY	SYRACUSE	WTKW	9
NY	ROCHESTER	WZXV	9
RI	PROVIDENCE	WXEX	10

99.7
CT	BRIDGEPORT	WEZN	2
MA	CAPE COD	WQRC	4
ME	PORTLAND	WMWX	5

100.1
MA	WORCESTER	WQVR	4

100.3
ME	BANGOR	WKIT	5
NH	PORTSMOUTH	WHEB	5
NY	NEW YORK CITY	WHTZ	8
RI	PROVIDENCE	WHKK	10

100.5
CT	HARTFORD	WRCH	2
NY	BINGHAMTON	WCDW	6
NY	ROCHESTER	WVOR	9

100.7
MA	BOSTON	WZLX	3
NY	NEW YORK CITY	WHUD	8
NY	UTICA	WVVC	10
NY	WATERTOWN	WOTT	10

100.9
CT	NEW LONDON	WTYD	2
MA	SPRINGFIELD	WRNX	4
ME	PORTLAND	WYNZ	5
NY	ALBANY	WKLI	5
NY	ELMIRA	WPGI	7
NY	SYRACUSE	WKRL	9

101.1
MA	CAPE COD	WWKJ	4
NH	MANCHESTER	WGIR	5
NY	ALBANY	WBUG	6
NY	NEW YORK CITY	WCBS	8

101.3
CT	NEW HAVEN	WKCI	2
ME	AUGUSTA	WKCG	4
NY	ALBANY	WJKE	5
NY	ROCHESTER	WRMM	9
NY	UTICA	WBRV	10

101.5
NY	POUGHKEEPSIE	WPDH	8
RI	PROVIDENCE	WWBB	10

101.7
MA	BOSTON	WFNX	3

NY	BINGHAMTON	WLTB	6
NY	LONG ISLAND	WBAZ	7
NY	SYRACUSE	WSCP	9

101.9

MA	CAPE COD	WCIB	4
ME	PORTLAND	WPOR	5
NY	NEW YORK CITY	WQCD	8

102.1

MA	SPRINGFIELD	WAQY	4
NH	PORTSMOUTH	WXBP	5
NY	BINGHAMTON	WAVR	6
NY	SYRACUSE	WRDS	9

102.3

CT	NEW LONDON	WVVE	2
NY	ALBANY	WXCR	6
NY	LONG ISLAND	WBAB	7
NY	ROCHESTER	WISY	9

102.2

MA	BOSTON	WCRB	3
NY	BUFFALO	WMJQ	6
NY	UTICA	WSKS	10

102.7

NY	NEW YORK CITY	WNEW	8
NY	ROCHESTER	WDCZ	9
RI	PROVIDENCE	WAKX	10

102.9

CT	HARTFORD	WDRC	2
MA	CAPE COD	WPXC	4
ME	PORTLAND	WBLM	5

103.1

NY	ALBANY	WHRL	6
NY	LONG ISLAND	WBZO	7
NY	NEWBURGH	WGNY	8
NY	WATERTOWN	WTOJ	10

103.3

MA	BOSTON	WODS	3
NY	BINGHAMTON	WMXW	6
NY	BUFFALO	WEDG	6

103.5

NY	ALBANY	WQBJ	6
NY	LONG ISLAND	WKTU	7
NY	ROCHESTER	WNNR	9

103.7

NY	ITHACA	WQNY	7
RI	PROVIDENCE	WWRX	10

103.9

MA	CAPE COD	WOCN	4
ME	BANGOR	WVOM	5
NY	ALBANY	WQBK	6
NY	LONG ISLAND	WRCN	7
NY	NEW YORK CITY	WFAS	8
NY	ROCHESTER	WDKX	9

104.1

CT	WATERBURY	WMRQ	3
MA	BOSTON	WBCN	3
NY	BINGHAMTON	WYOS	6
NY	BUFFALO	WHTT	6

104.3

ME	AUGUSTA	WABK	4
NY	NEW YORK CITY	WAXQ	8
NY	UTICA	WFRG	10

104.5

MA	WORCESTER	WXLO	4
NY	ALBANY	WXLE	5

104.7

MA	CAPE COD	WKPE	4
NY	POUGHKEEPSIE	WSPK	8
NY	SYRACUSE	WBBS	9

104.9

MA	BOSTON	WBOQ	3

105.1

NY	NEW YORK CITY	WBIX	8
RI	PROVIDENCE	WWLI	10

105.3

NH	PORTSMOUTH	WXBB	5
NY	ELMIRA	WKPQ	7

105.5

CT	NEW LONDON	WQGN	2
NY	SYRACUSE	WTKV	9
NY	UTICA	WOWB	10

105.7

MA	BOSTON	WROR	3
NY	BINGHAMTON	WMRV	6

105.9

CT	HARTFORD	WHCN	2
NY	NEW YORK CITY	WCAA	8
NY	ROCHESTER	WJZR	9
NY	SYRACUSE	WLTI	9

106.1

MA	CAPE COD	WCOD	4
NY	ELMIRA	WNKI	7
NY	LONG ISLAND	WBLI	7

106.3

ME	PORTLAND	WBQW	5
NY	SYRACUSE	WMCR	9
RI	PROVIDENCE	WWKX	10

106.5

CT	NEW LONDON	WBMW	2
ME	BANGOR	WQCB	5
NY	ALBANY	WPYX	6
NY	BUFFALO	WYRK	6

106.7

MA	BOSTON	WMJX	3
ME	PORTLAND	WLAM	5
NY	NEW YORK CITY	WLTW	8
NY	ROCHESTER	WYSY	9
NY	WATERTOWN	WWLF	10

106.9

CT	HARTFORD	WCCC	2

107.1

NH	PORTSMOUTH	WERZ	5
NY	NEW YORK CITY	WWXY	8
RI	PROVIDENCE	WFHN	10

107.3

ME	BANGOR	WBZN	5
NY	ROCHESTER	WMAX	9
NY	UTICA	WRCK	10

107.5

MA	CAPE COD	WFCC	4
ME	PORTLAND	WTHT	5
NY	NEW YORK CITY	WBLS	8

107.7

CT	NEW LONDON	WKCD	2
NY	ALBANY	WGNA	6

107.9

CT	STAMFORD	WEBE	2
MA	BOSTON	WXKS	3
NY	SYRACUSE	WWHT	9
VT	BURLINGTON	WVPS	11

AM

540

NY	LONG ISLAND	WLUX	7

550

NY	BUFFALO	WGR	7
RI	PROVIDENCE	WLKW	10

560

MA	SPRINGFIELD	WHYN	4
ME	PORTLAND	WGAN	5

570

NY	NEW YORK CITY	WMCA	8
NY	SYRACUSE	WSYR	10

580

MA	WORCESTER	WTAG	4

590

MA	BOSTON	WEZE	3
NY	ALBANY	WROW	6

600

CT	BRIDGEPORT	WICC	2

610

NH	MANCHESTER	WGIR	5

620

ME	BANGOR	WZON	5
NY	NEW YORK CITY	WJWR	8
NY	SYRACUSE	WHEN	10
VT	BURLINGTON	WVMT	11

630

RI	PROVIDENCE	WPRO	11

640

MA	SPRINGFIELD	WNNZ	4

660

NY	NEW YORK CITY	WFAN	8

680

MA	BOSTON	WRKO	3
NY	BINGHAMTON	WINR	6

710			
NY	NEW YORK CITY	WOR	8

730			
MA	SPRINGFIELD	WACE	4

770			
NY	NEW YORK CITY	WABC	8

790			
NY	WATERTOWN	WTNY	10

800			
CT	DANBURY	WLAD	2
MA	BOSTON	WCCM	3

810			
NY	ALBANY	WGY	6

820			
NY	ELMIRA	WWLZ	7
NY	NEW YORK CITY	WNYC	8

840			
CT	HARTFORD	WRYM	2

850			
CT	DANBURY	WREF	2
MA	BOSTON	WEEI	3

870			
ME	PORTLAND	WLAM	5
NY	ITHACA	WHCU	7

880			
NY	NEW YORK CITY	WCBS	8

910			
CT	HARTFORD	WNEZ	2
ME	BANGOR	WABI	5

920			
RI	PROVIDENCE	WHJJ	11

930			
NH	PORTSMOUTH	WZNN	5
NY	BUFFALO	WBEN	7
NY	NEW YORK CITY	WPAT	8

950			
MA	BOSTON	WROL	3
NY	ROCHESTER	WEZO	9
NY	UTICA	WIBX	10

960			
CT	NEW HAVEN	WELI	2

970			
ME	PORTLAND	WZAN	5
NY	BUFFALO	WNED	7
NY	NEW YORK CITY	WWDJ	8

980			
MA	BOSTON	WCAP	3
NY	ALBANY	WTRY	6

990			
RI	PROVIDENCE	WALE	11

1010			
NY	NEW YORK CITY	WINS	8

1030			
MA	BOSTON	WBZ	3

1040			
NY	ROCHESTER	WYSL	9

1050			
NY	NEW YORK CITY	WEVD	8

1080			
CT	HARTFORD	WTIC	2
NY	BUFFALO	WUFO	7

1120			
CT	HARTFORD	WPRX	2

1130			
NY	NEW YORK CITY	WBBR	8

1150			
NY	UTICA	WRUN	10

1180			
NY	ROCHESTER	WHAM	9

1200			
NY	SYRACUSE	WTLA	9

1220			
CT	NEW HAVEN	WQUN	2
NY	NEWBURGH	WGNY	8

1230			
CT	HARTFORD	WLAT	2
MA	BOSTON	WESX	3
NY	BUFFALO	WECK	6
NY	ELMIRA	WENY	7
NY	UTICA	WLFH	10
RI	PROVIDENCE	WERI	11
VT	BURLINGTON	WJOY	11

1240			
CT	WATERBURY	WWCO	3
MA	CAPE COD	WBUR	4
NY	WATERTOWN	WATN	10
RI	PROVIDENCE	WOON	10

1250			
MA	SPRINGFIELD	WARE	4
NH	MANCHESTER	WKBR	5

1260			
MA	BOSTON	WPZE	3
NY	POUGHKEEPSIE	WBNR	9
NY	SYRACUSE	WNSS	10

1270			
MA	SPRINGFIELD	WSPR	4
NH	PORTSMOUTH	WTSN	5
NY	BUFFALO	WHLD	7

1280			
ME	AUGUSTA	WFAU	4
NY	NEW YORK CITY	WADO	8
NY	ROCHESTER	WHTK	9

1290			
NY	BINGHAMTON	WNBF	6
RI	PROVIDENCE	WRNI	11

1300			
CT	NEW HAVEN	WAVZ	2
NY	ALBANY	WTMM	6
NY	BUFFALO	WXRL	6

1310			
CT	NEW LONDON	WICH	2
MA	WORCESTER	WORC	4
NY	UTICA	WTLB	10

1320			
CT	WATERBURY	WATR	3
NH	MANCHESTER	WDER	5
NY	ELMIRA	WHHO	7
RI	PROVIDENCE	WJYT	10

1330			
MA	BOSTON	WRCA	3

1340			
NH	PORTSMOUTH	WWNH	5
NY	BUFFALO	WLVL	7
NY	NEWBURGH	WALL	8
RI	PROVIDENCE	WNBH	10

1350			
CT	STAMFORD	WNLK	3
NY	ELMIRA	WCBA	7

1360			
CT	HARTFORD	WDRC	2
NY	BINGHAMTON	WKOP	6

1370			
NH	MANCHESTER	WFEA	5
NY	LONG ISLAND	WALK	7
NY	ROCHESTER	WXXI	9

1380			
NY	ELMIRA	WABH	7

1390			
MA	BOSTON	WPLM	3
NY	POUGHKEEPSIE	WEOK	9
NY	SYRACUSE	WDCW	10
VT	BURLINGTON	WKDR	11

1400			
CT	STAMFORD	WSTC	3
MA	BOSTON	WLLH	3
MA	SPRINGFIELD	WHMP	4
NY	ALBANY	WABY	6
NY	BUFFALO	WWWS	7
RI	PROVIDENCE	WHTB	11

1410			
CT	HARTFORD	WPOP	2
NY	ELMIRA	WELM	7
NY	WATERTOWN	WUZZ	10

1420			
RI	PROVIDENCE	WBSM	11

1430			
MA	BOSTON	WXKS	3
NY	BINGHAMTON	WENE	6

1440			
MA	WORCESTER	WWTM	4
ME	PORTLAND	WJAE	5

1450			
CT	BRIDGEPORT	WCUM	3
MA	SPRINGFIELD	WMAS	4
NY	ELMIRA	WCLI	7

NY	POUGHKEEPSIE	WKIP	9
NY	UTICA	WODZ	10
RI	PROVIDENCE	WDYZ	11

1460

MA	BOSTON	WBET	3
NY	ALBANY	WGNA	6
NY	ROCHESTER	WWWG	9

1470

CT	HARTFORD	WMMW	2
NY	ITHACA	WTKO	7

1480

NY	UTICA	WADR	10
RI	PROVIDENCE	WSAR	11

1490

CT	STAMFORD	WGCH	2
MA	SPRINGFIELD	WACM	4
NY	ALBANY	WCSS	6
NY	ROCHESTER	WBTA	9
NY	SYRACUSE	WOLF	10

1500

CT	BRIDGEPORT	WFIF	2

1510

MA	BOSTON	WNRB	3

1520

NY	BUFFALO	WWKB	7

1540

NH	PORTSMOUTH	WMYF	5
NY	ALBANY	WDCD	6
NY	SYRACUSE	WSIV	10

1550

CT	HARTFORD	WDZK	2
NY	UTICA	WUTQ	10

1560

NY	NEW YORK CITY	WQEW	8

1570

NY	ALBANY	WBUG	6
RI	PROVIDENCE	WPEP	11

1590

NY	ELMIRA	WEHH	7
RI	PROVIDENCE	WARV	11

1600

NY	NEW YORK CITY	WWRL	8

MICHIGAN

ANN ARBOR	25
BATTLE CREEK	25
DETROIT	25
FLINT	25
GRAND RAPIDS	26
KALAMAZOO	26
LANSING	26
NORTHWEST MICH.	26
SAGINAW	27

WISCONSIN

APPLETON-OSHKOSH	37
EAU CLAIRE	37
GREEN BAY	37
LA CROSSE	37
MADISON	38
MILWAUKEE	38
WAUSAU	38

PENNSYLVANIA

ALLENTOWN-BETHL	31
ALTOONA	31
ERIE	31
HARRISBURG	31
JOHNSTOWN	32
LANCASTER	32
PHILADELPHIA	32
PITTSBURGH	33
READING	33
SCRANTON	34
STATE COLLEGE	34
WILLIAMSPORT	34
YORK	35

NEW JERSEY

ATLANTIC CITY	27
MONMOUTH-OCEAN CTY.	27
MORRISTOWN	27
TRENTON	27

MARYLAND

BALTIMORE	23
FREDERICK	24
HAGERSTOWN	24
SALISBURY	24

B

MICHIGAN

WISCONSIN

PENNSYLVANIA

NEW JERSEY

OHIO

DELAWARE

ILLINOIS

INDIANA

DC

MARYLAND

WEST VIRGINIA

VIRGINIA

KENTUCKY

0   100   200 Miles

WASHINGTON D.C.

WASHINGTON D.C.	18

ILLINOIS

BLOOMINGTON	18
CHAMPAIGN	18
CHICAGO	19
DANVILLE	19
MARION-CARBOND	20
PEORIA	20
ROCKFORD	20
SPRINGFIELD	20

INDIANA

BLOOMINGTON	20
EVANSVILLE	21
FT. WAYNE	21
INDIANAPOLIS	21
LAFAYETTE	22
SOUTH BEND	22
TERRE HAUTE	22

KENTUCKY

LEXINGTON	22
LOUISVILLE	23
OWENSBORO	23

OHIO

AKRON	28
CANTON	28
CINCINNATI	28
CLEVELAND	29
COLUMBUS	29
DAYTON	30
LIMA	30
TOLEDO	30
YOUNGSTOWN	30

WEST VIRGINIA

BECKLEY	38
CHARLESTON	39
HUNTINGTON	39
MORGANTOWN	39
PARKERSBURG	39
WHEELING	40

VIRGINIA

CHARLOTTESVILLE	35
HARRISONBURG	35
NORFOLK	36
RICHMOND	36
ROANOKE	36

DC
DE
IL

FREQ	STATION	POWER	FORMAT	
97.1	WASH	IIIII	LGHT AC	
99.5	WGAY	III	LGHT AC	
107.3	WRQX	IIIII	AC	
104.1	WWZZ	IIIIIII	TOP 40	
94.7	WARW	IIIIIIII	CLS RK	
101.1	WWDC	IIIIIIII	ROCK	
100.3	WBIG	III	OLDS	
98.7	WMZQ	IIIII	CTRY	
AM 1540	WACA	IIIII	SPAN	
103.5	WGMS	IIIII	CLS	
105.9	WJZW	IIIII	JZZ	
AM 1260	WWDC	IIIII	BB, NOS	
88.1	WMUC	III	DVRS	C
88.5	WAMU	IIIII	NWS, TLK, NPR, PRI	P
89.3	WPFW	III	JZZ	P
90.9	WETA	IIIIIIII	CLS, NPR, PRI	P
AM 1340	WYCB	I	GOSP	
AM 1580	WPGC	IIIIIIII	GOSP	
91.9	WGTS	III	RLG AC	C
92.7	WMJS	III	EZ	
105.1	WAVA	IIIIIIII	RLG	
93.9	WKYS	III	URBAN	
95.5	WPGC	IIIII	URBAN	
96.3	WHUR	III	URBAN	
102.3	WMMJ	III	URBAN	
106.7	WJFK	III	NWS, TLK	
AM 570	WWRC	IIIIIIII	BUS NWS	
AM 630	WMAL	IIIIIIII	NWS, TLK	
AM 980	WTEM	IIIIIIII	SPRTS	
AM 1390	WZHF	IIIII	NWS, TLK	
AM 1450	WOL	I	NWS, TLK	
AM 1500	WTOP	IIIIIIII	NWS	

WILMINGTON, DE

FREQ	STATION	POWER	FORMAT	
93.7	WSTW	IIIII	AC	
99.5	WJBR	IIIII	AC	
101.7	WJKS	III	TOP 40	
91.3	WVUD	I	DVRS	C
AM 1150	WDEL	IIIII	NWS, TLK	
AM 1450	WILM	I	NWS	

BLOOMINGTON, IL

FREQ	STATION	POWER	FORMAT	
101.5	WBNQ	IIIII	TOP 40	
96.7	WIHN	III	ROCK	
104.1	WBWN	III	CTRY	
88.1	WESN	I	DVRS	C
89.1	WGLT	I	DVRS, NPR	C
AM 1230	WJBC	I	FS	

CHAMPAIGN, IL

FREQ	STATION	POWER	FORMAT	
94.5	WLRW	IIIII	AC	
96.1	WQQB	III	AC	
97.5	WHMS	IIIII	AC	
95.3	WZNF	III	ROCK	
105.9	WGKC	III	CLS RK	
107.1	WPGU	I	MOD RK	
92.5	WKIO	III	OLDS	
100.3	WIXY	III	CTRY	
88.7	WPCD	III	OLDS	C
90.1	WEFT	III	DVRS	P
90.9	WILL	IIIIIIII	CLS, NPR, PRI	P
AM 580	WILL	IIIIIIII	NWS, TLK, NPR, PRI	P

IL

	FREQ	STATION	POWER	FORMAT	
AM	1400	WDWS	I	NWS, TLK	
		CHICAGO, IL			
	93.9	WLIT	III	AC	
	100.3	WNND	III	AC	
	101.9	WTMX	III	AC	
	102.7	WVAZ	III	AC	
	105.5	WZSR	III	AC	
	96.3	WBBM	III	TOP 40	
	102.3	WXLC	III	TOP 40	
	93.1	WXRT	III	ALT RK	
	94.7	WXCD	III	CLS RK	
	95.1	WIIL	IIIII	CLS RK	
	97.9	WLUP	IIIIIIII	CLS RK	
	101.1	WKQX	III	DVRS	
	103.5	WRCX	III	ROCK	
	104.3	WJMK	IIIIIIII	OLDS	
AM	1390	WGCI	IIIII	OLDS	
	98.3	WCCQ	III	CTRY	
	99.5	WUSN	III	CTRY	
	105.5	WLJE	III	CTRY	
	103.9	WZCH	III	SPAN	
	105.1	WOJO	III	SPAN	
	107.9	WLEY	III	SPAN	
AM	560	WIND	IIIIIIII	SPAN	
AM	1200	WLXX	IIIII	SPAN	
	107.5	WGCI	IIIII	BLK, R&B	
	95.5	WNUA	IIIII	JZZ	
	96.9	WNIZ	IIIII	CLS	
	97.1	WNIB	III	CLS	
	98.7	WFMT	IIIII	CLS	
AM	850	WAIT	IIIII	BB, NOS	
	88.1	WETN	I	DVRS	C
	88.1	WLRA	I	DVRS	C
	88.1	WCRX	I	DVRS	C
	88.3	WZRD	I	DVRS	C
	88.7	WRSE	I	ROCK	C
	88.7	WLUW	I	TOP 40	C
	88.9	WMXM	I	AC	C
	88.9	WRRG	I	TOP 40	C
	89.1	WONC	I	ROCK	C
	89.3	WNUR	I	JZZ	C
	90.9	WDCB	III	JZZ, CLS, PRI	C
	91.5	WBEZ	III	NWS, TLK, NPR, PRI	P
	92.3	WYCA	IIIII	GOSP	
	89.3	WKKC	I	EDU	C
	106.3	WYBA	III	RLG	
	106.7	WYLL	IIIII	RLG	
AM	1300	WTAQ	IIIII	KIDS	
	105.9	WCKG	III	NWS, TLK	
AM	670	WMAQ	IIIIIIII	NWS	
AM	720	WGN	IIIIIIII	NWS, TLK	
AM	780	WBBM	IIIIIIII	NWS	
AM	890	WLS	IIIIIIII	NWS, TLK	
AM	950	WIDB	IIIIIIII	SPRTS	
AM	1000	WMVP	IIIIIIII	NWS, TLK	
AM	1160	WSCR	IIIIIIII	SPRTS	
AM	1230	WJOB	I	NWS, TLK	
AM	1450	WVON	I	NWS, TLK	
		DANVILLE, IL			
	102.1	WDNL	IIIII	AC	
	94.9	WRHK	III	CLS RK	
	99.1	WIAI	IIIII	CTRY	

19

	FREQ	STATION	POWER	FORMAT	
	100.9	WHPO	III	CTRY	
AM	980	WITY	III	BB, NOS	
AM	1490	WDAN	I	NWS, TLK	

CARBONDALE, IL

	FREQ	STATION	POWER	FORMAT	
	92.7	WVZA	III	AC	
	103.5	WUEZ	III	AC	
	101.5	WCIL	IIIII	TOP 40	
	95.1	WXLT	III	CLS RK	
	97.7	WQUL	III	CLS RK	
	105.1	WTAO	III	CLS RK	
	106.3	WQRL	III	OLDS	
	99.9	WOOZ	IIIIIII	CTRY	
	107.3	WDDD	IIIII	CTRY	
	91.9	WSIU	IIIII	CLS, NPR, PRI	P
	103.9	WXAN	III	GOSP	
AM	810	WDDD	III	SPRTS	
AM	1340	WJPF	I	NWS, TLK	
AM	1420	WINI	I	NWS, TLK	

PEORIA, IL

	FREQ	STATION	POWER	FORMAT	
	98.5	WEEK	III	AC	
	106.9	WSWT	IIIII	LGHT AC	
	94.3	WFXF	III	CLS RK	
	95.5	WGLO	IIIII	ROCK	
	99.9	WIXO	III	MOD RK	
	105.7	WWCT	III	ROCK	
	93.3	WPBG	IIIIIII	OLDS	
	97.3	WFYR	III	CTRY	
	104.9	WXCL	III	CTRY	
AM	1350	WOAM	I	BB, NOS	
	89.9	WCBU	IIIII	CLS, NPR, PRI	P
	92.3	WBGE	I	URBAN	
	102.3	WTAZ	III	NWS, TLK	
AM	1290	WIRL	IIIII	NWS, TLK	
AM	1470	WMBD	IIIII	NWS, TLK	

ROCKFORD, IL

	FREQ	STATION	POWER	FORMAT	
	103.1	WRWC	III	AC	
	97.5	WZOK	IIIII	TOP 40	
	104.9	WXRX	III	ROCK	
	95.3	WKMQ	III	OLDS	
	96.7	WLUV	I	CTRY	
	90.5	WNIU	III	CLS, NPR, PRI	C
	100.9	WQFL	III	RLG AC	
AM	1380	WTJK	IIIII	NWS, TLK	
AM	1440	WROK	IIIII	NWS, TLK	

SPRINGFIELD, IL

	FREQ	STATION	POWER	FORMAT	
	98.7	WNNS	IIIII	AC	
	103.7	WDBR	IIIII	TOP 40	
	101.9	WQQL	IIIII	OLDS	
	104.5	WFMB	IIIII	CTRY	
	91.9	WUIS	IIIII	CLS, NPR, PRI	P
AM	970	WMAY	III	NWS, TLK	
AM	1240	WTAX	I	NWS, TLK	
AM	1450	WFMB	I	SPRTS	

BLOOMINGTON, IN

	FREQ	STATION	POWER	FORMAT	
	96.7	WBWB	III	TOP 40	
	92.3	WTTS	IIIIIII	ALT RK	
	91.3	WFHB	III	DVRS	P

IL
IN

	FREQ	STATION	POWER	FORMAT	
	103.7	WFIU	IIIII	CLS, NPR, PRI	P
	105.1	WGCT	I	CTRY	C
AM	1370	WGCL	IIIII	NWS, TLK	

EVANSVILLE, IN

	FREQ	STATION	POWER	FORMAT	
	104.1	WIKY	IIIII	AC	
	106.1	WDKS	III	HOT AC	
	103.1	WGBF	III	ROCK	
	107.5	WABX	I	CLS RK	
	93.5	WJPS	III	OLDS	
	99.5	WKDQ	IIIIIIII	CTRY	
	105.3	WYNG	IIIII	CTRY	
AM	860	WSON	III	BB, NOS	
	88.3	WNIN	IIIII	CLS, NPR, PRI	P
	90.9	WKUE	IIIII	CLS, NPR, PRI	P
	91.5	WUEV	I	DVRS	C
AM	1330	WVHI	IIIII	RLG AC	
	107.1	WBNL	I	EZ	
AM	1280	WGBF	IIIII	NWS, TLK	
AM	1400	WEOA	I	URBAN	

FT. WAYNE, IN

	FREQ	STATION	POWER	FORMAT	
	95.1	WAJI	IIIII	AC	
	97.3	WMEE	IIIIIIII	HOT AC	
	101.1	WLZQ	III	HOT AC	
	102.3	WGL	III	AC	
	92.3	WFWI	III	CLS RK	
	94.1	WYSR	III	MOD RK	
	96.3	WEJE	III	DVRS	
	102.9	WEXI	III	ROCK	
	103.9	WXKE	III	ROCK	
	101.7	WLDE	III	OLDS	
	105.1	WQHK	III	CTRY	
	106.3	WSHI	III	BB, NOS	
	89.1	WBNI	III	CLS, NPR, PRI	P
	107.9	WJFX	III	URBAN	
AM	1190	WOWO	IIIIIIII	NWS, TLK	
AM	1250	WGL	III	NWS, TLK	
AM	1380	WHWD	IIIII	SPRTS	
AM	1570	WGLL	I	NWS, TLK	

INDIANAPOLIS, IN

	FREQ	STATION	POWER	FORMAT	
	97.1	WENS	III	AC	
	107.9	WTPI	III	AC	
	96.3	WHHH	III	TOP 40	
	99.5	WZPL	III	TOP 40	
	93.1	WNAP	IIIII	CLS RK	
	94.7	WFBQ	IIIIIIII	CLS RK	
	103.3	WRZX	III	MOD RK	
	101.9	WQFE	III	OLDS	
	104.5	WGLD	III	OLDS	
	93.9	WGRL	III	CTRY	
	95.5	WFMS	IIIIIIII	CTRY	
	102.3	WCBK	III	CTRY	
	105.7	WTLC	IIIII	BLK, R&B	
AM	1310	WTLC	IIIII	BLK, R&B	
	100.9	WYJZ	III	JZZ	
	107.1	WSYW	III	JZZ	
AM	1430	WMYS	IIIII	BB, NOS	
	88.7	WICR	III	CLS, NPR, PRI	P
	89.5	WFCI	I	AC	C
	90.1	WFYI	III	NWS, TLK, NPR, PRI	P

**IN
KY**

	FREQ	STATION	POWER	FORMAT	
	95.9	WPZZ	III	GOSP	
	98.3	WXIR	III	RLG AC	
AM	950	WXLW	IIIIIIII	RLG AC	
AM	1500	WBRI	IIIII	RLG AC	
AM	1590	WNTS	IIIII	GOSP	
	106.7	WBKS	III	URBAN	
AM	1070	WIBC	IIIIIIII	NWS, TLK	
AM	1260	WNDE	IIIII	SPRTS	

LAFAYETTE, IN

	FREQ	STATION	POWER	FORMAT	
	96.5	WAZY	IIIII	HOT AC	
	106.7	WGLM	III	AC	
	93.5	WKHY	III	CLS RK	
	98.7	WASK	III	OLDS	
AM	1450	WASK	I	OLDS	
	105.3	WKOA	IIIII	CTRY	
	101.3	WBAA	III	CLS, NPR, PRI	P
AM	920	WBAA	IIIIIIII	NWS, TLK, NPR, PRI	C

SO. BEND, IN

	FREQ	STATION	POWER	FORMAT	
	101.5	WNSN	III	AC	
	103.9	WRBR	III	AC	
	92.9	WNDU	III	TOP 40	
AM	1490	WNDU	I	70's RK	
AM	1580	WHLY	I	BB, NOS	
	88.9	WSND	III	CLS	C
	102.3	WGTC	III	RLG AC	
	103.1	WHME	III	RLG	
	106.3	WUBU	III	URBAN	
AM	960	WSBT	IIIIIIII	NWS, TLK	

TERRE HAUTE, IN

	FREQ	STATION	POWER	FORMAT	
	100.7	WMGI	IIIII	AC	
	104.3	WCBH	III	HOT AC	
	105.9	WMMC	III	AC	
	105.5	WWVR	III	CLS RK	
	107.5	WZZQ	IIIII	ROCK	
	97.7	WSDM	III	OLDS	
AM	1440	WPRS	I	OLDS	
	95.3	WNDI	I	CTRY	
	98.5	WACF	IIIII	CTRY	
	99.9	WTHI	IIIII	CTRY	
	89.7	WISU	III	JZZ	C
	90.5	WMHD	I	DVRS	C
	102.7	WLEZ	IIIIIIII	EZ	
AM	640	WBOW	IIIII	NWS, TLK	
AM	1480	WTHI	IIIII	NWS, TLK	

LEXINGTON, KY

	FREQ	STATION	POWER	FORMAT	
	94.5	WMXL	IIIIIIII	HOT AC	
	96.9	WGKS	IIIII	AC	
	106.7	WKXO	III	AC	
AM	590	WVLK	IIIIIIII	AC	
AM	1340	WEKY	I	AC	
	104.5	WLKT	IIIII	TOP 40	
	100.1	WKQQ	IIIIIIII	CLS RK	
	101.5	WLRO	IIIII	CLS RK	
	103.3	WXZZ	III	MOD RK	
	102.5	WLTO	III	OLDS	

BOSTON, MA

	FREQ	STATION	POWER	FORMAT	
	88.1	WRFL	III	DVRS	C

	FREQ	STATION	POWER	FORMAT	
	88.9	WEKU	IIIIIII	CLS, NPR, PRI	P
	89.9	WRVG	I	DVRS	C
	91.3	WUKY	IIIIIII	JZZ, NPR, PRI	P
	105.9	WVRB	III	RLG AC	
	106.3	WJMM	III	RLG	
AM	630	WLAP	IIIIIII	NWS, TLK	
AM	1300	WLXG	III	NWS, TLK	

LOUISVILLE, KY

	FREQ	STATION	POWER	FORMAT	
	103.9	WMHX	III	AC	
	105.9	WRVI	III	AC	
	106.9	WVEZ	IIIII	AC	
	99.7	WDJX	IIIIIII	TOP 40	
	95.7	WQMF	IIIII	CLS RK	
	100.5	WTFX	IIIIIII	ROCK	
	102.3	WLRS	III	DVRS	
	107.7	WSFR	III	CLS RK	
	103.1	WRKA	III	OLDS	
	97.5	WAMZ	IIIII	CTRY	
	101.7	WTHQ	III	CTRY	
	105.3	WMPI	III	CTRY	
	94.7	WLSY	III	BLK, R&B	
AM	1080	WKJK	IIIII	BB, NOS	
AM	1450	WAVG	I	BB, NOS	
	89.3	WFPL	III	NWS, TLK, NPR, PRI	P
	90.5	WUOL	IIIII	CLS, NPR, PRI	P
	91.9	WFPK	IIIII	ALT RK, NPR, PRI	P
	105.1	WXLN	I	RLG AC	
AM	1240	WLLV	I	GOSP	
AM	1350	WLOU	IIIII	GOSP	
AM	900	WFIA	III	RLG	
	96.5	WGZB	I	URBAN	
	101.3	WMJM	I	URBAN	
AM	620	WTMT	IIIII	SPRTS	
AM	680	WNAI	IIIII	NWS, TLK	
AM	790	WWKY	IIIIIII	NWS, TLK	
AM	840	WHAS	IIIIIII	NWS, TLK	
AM	970	WLKY	IIIIIII	NWS	

OWENSBORO, KY

	FREQ	STATION	POWER	FORMAT	
	92.5	WBKR	IIIIIII	CTRY	
	94.7	WBIO	III	CTRY	
AM	1420	WVJS	IIIII	BB, NOS	
	90.3	WKWC	I	CLS	P
AM	1490	WOMI	I	NWS, TLK	

BALTIMORE, MD

	FREQ	STATION	POWER	FORMAT	
	101.9	WLIF	IIIIIII	AC	
	104.3	WOCT	IIIII	AC	
	106.5	WWMX	IIIIIII	AC	
	102.7	WXYV	IIIII	TOP 40	
	97.9	WIYY	IIIIIII	ROCK	
	103.1	WRNR	III	DVRS	
	105.7	WQSR	IIIII	OLDS	
	93.1	WPOC	IIIIIII	CTRY	
	100.7	WGRX	IIIIIII	CTRY	
	103.7	WXCY	IIIII	CTRY	
AM	1330	WASA	IIIII	BB, NOS	
AM	1360	WWLG	IIIII	BB, NOS	
	88.1	WJHU	III	CLS, NPR, PRI	P
	88.9	WEAA	III	JZZ, NPR	P
	89.7	WTMD	III	AC	P

23

BALTIMORE • FREDERICK • HAGERSTOWN
SALISBURY/OCEAN CITY, MD

MD

	FREQ	STATION	POWER	FORMAT	
	91.5	WBJC	IIIII	CLS, NPR, PRI	P
AM	600	WCAO	IIIIIIII	RLG AC	
AM	860	WBGR	IIIII	GOSP	
AM	1400	WWIN	I	GOSP	
	95.1	WRBS	IIIII	RLG	
AM	1230	WITH	I	RLG	
	92.3	WERQ	IIIII	URBAN	
	95.9	WWIN	III	URBAN	
AM	680	WCBM	IIIIIIII	NWS, TLK	
AM	1090	WBAL	IIIIIIII	NWS, TLK	
AM	1300	WJFK	IIIII	NWS, TLK	

FREDERICK, MD

	FREQ	STATION	POWER	FORMAT	
	103.1	WAFY	III	AC	
	103.9	WWVZ	III	TOP 40	
	99.9	WFRE	III	CTRY	
AM	820	WXTR	IIIIIIII	CTRY	
	89.9	WMTB	I	DVRS	C
AM	930	WFMD	IIIIIIII	NWS, TLK	

HAGERSTOWN, MD

	FREQ	STATION	POWER	FORMAT	
	95.1	WIKZ	IIIII	AC	
AM	1240	WJEJ	I	LGHT AC	
	92.1	WSRT	III	CLS RK	
	96.7	WQCM	I	ROCK	
	106.9	WARX	III	OLDS	
AM	1490	WARK	I	OLDS	
	94.3	WCHA	III	CTRY	
	95.9	WYII	III	CTRY	
	101.5	WAYZ	III	CTRY	
AM	800	WCHA	III	CTRY	
	89.1	WETH	III	CLS, NPR, PRI	P
	104.7	WWMD	III	EZ	
AM	1590	WCBG	IIIII	NWS, TLK	

SALISBURY, MD

	FREQ	STATION	POWER	FORMAT	
	95.3	WJNE	III	AC	
	97.7	WAFL	III	AC	
	103.5	WJYN	III	AC	
	104.7	WQHQ	IIIIIIII	AC	
	103.9	WOCQ	III	TOP 40	
	106.9	WRXS	III	TOP 40	
	93.5	WZBH	IIIII	ROCK	
	95.9	WOSC	III	DVRS	
	98.5	WGBG	III	CLS RK	
	101.7	WRBG	III	CLS RK	
	92.1	WLBW	III	OLDS	
	105.5	WLVW	III	OLDS	
	106.5	WKHW	III	OLDS	
	94.3	WICO	III	CTRY	
	96.9	WBEY	III	CTRY	
	97.9	WSBL	III	CTRY	
	99.9	WWFG	IIIII	CTRY	
	105.9	WXJN	III	CTRY	
	97.1	WQJZ	III	JZZ	
AM	900	WJWL	IIIIIIII	BB, NOS	
	89.5	WSCL	IIIII	CLS, NPR, PRI	P
	91.3	WESM	IIIII	JZZ, NPR, PRI	P
	101.3	WXPZ	III	RLG AC	
	102.5	WOLC	IIIII	RLG AC	
	92.7	WGMD	III	NWS, TLK	
AM	960	WTGM	IIIIIIII	SPRTS	

24

	FREQ	STATION	POWER	FORMAT	
	102.9	WIQB	IIIII	ROCK	
	107.1	WQKL	III	OLDS	
AM	1600	WAAM	IIIII	BB, NOS	
	88.3	WCBN	I	DVRS	C
	89.1	WEMU	III	JZZ, NPR	P
	91.7	WUOM	IIIIIIII	NWS, TLK, NPR, PRI	P
AM	1050	WTKA	IIIII	SPRTS	

BATTLE CREEK, MI

	FREQ	STATION	POWER	FORMAT	
	95.3	WBXX	III	HOT AC	
	104.9	WWKN	III	OLDS	
AM	1400	WRCC	I	BB, NOS	
	96.7	WUFN	III	RLG	P
AM	1500	WOLY	I	RLG	
AM	930	WBCK	IIIIIIII	NWS, TLK	

DETROIT, MI

	FREQ	STATION	POWER	FORMAT	
	92.3	WMXD	III	AC	
	93.1	WDRQ	IIIII	AC	
	93.5	WHMI	III	AC	
	95.5	WKQI	IIIII	HOT AC	
	96.3	WPLT	IIIIIIII	AC	
	100.3	WNIC	IIIII	AC	
AM	1380	WPHM	IIIII	AC	
	94.7	WCSX	IIIIIIII	CLS RK	
	97.1	WKRK	III	ROCK	
	101.1	WRIF	IIIII	ROCK	
	102.7	WWBR	IIIII	CLS RK	
	105.1	WXDG	IIIIIIII	DVRS	
	104.3	WOMC	IIIII	OLDS	
	99.5	WYCD	III	CTRY	
	106.7	WWWW	IIIIIIII	CTRY	
	107.1	WSAQ	III	CTRY	
AM	1400	WQBH	I	BLK, R&B	
	98.7	WVMV	IIIII	JZZ	
	89.3	WHFR	I	DVRS	C
	91.3	WSGR	I	DVRS	C
	101.9	WDET	IIIII	DVRS, NPR, PRI	P
	103.5	WMUZ	IIIII	RLG AC	
AM	690	WNZK	IIIIIIII	ETH	
	97.9	WJLB	IIIII	URBAN	
	105.9	WCHB	III	URBAN	
	107.5	WGPR	IIIII	URBAN	
AM	760	WJR	IIIIIIII	NWS, TLK	
AM	950	WWJ	IIIIIIII	NWS	
AM	1130	WDFN	IIIIIIII	SPRTS	
AM	1270	WXYT	IIIII	NWS, TLK	
AM	1310	WYUR	IIIII	NWS, TLK	

FLINT, MI

	FREQ	STATION	POWER	FORMAT	
	107.9	WCRZ	IIIII	AC	
	105.5	WWCK	III	TOP 40	
	95.1	WFBE	III	CTRY	
	91.1	WFUM	III	NWS, TLK, NPR, PRI	P
	95.9	WGRI	I	GOSP	P
AM	600	WSNL	IIIII	RLG AC	
AM	1160	WWON	I	GOSP	
AM	1420	WFLT	I	RLG	
	92.7	WDZZ	III	URBAN	
AM	910	WFDF	IIIIIIII	NWS, TLK	
AM	1330	WTRX	IIIII	SPRTS	
AM	1470	WFNT	IIIII	NWS, TLK	

MI

FREQ	STATION	POWER	FORMAT	
92.1	WGHN	I	AC	
95.7	WLHT	IIIII	AC	
100.5	WTRV	III	LGHT AC	
105.7	WOOD	IIIIIIII	LGHT AC	
96.1	WVTI	IIIII	TOP 40	
94.5	WKLQ	IIIII	MOD RK	
96.9	WLAV	IIIII	CLS RK	
97.9	WGRD	III	MOD RK	
93.7	WBCT	IIIIIIII	CTRY	
101.3	WCUZ	IIIII	CTRY	
98.7	WFGR	I	CLS	
88.5	WGVU	III	NWS, NPR, PRI	P
88.9	WBLU	I	CLS, NPR, PRI	P
89.9	WTHS	I	DVRS	C
99.3	WJQK	III	RLG AC	
102.9	WFUR	IIIIIIII	RLG AC	
104.1	WVGR	IIIIIIII	NWS, TLK, NPR, PRI	P
AM 1480	WGVU	III	NWS, TLK, NPR, PRI	P
AM 1260	WWJQ	IIIII	RLG	
AM 1450	WHTC	I	FS	
AM 1140	WKWM	IIIII	URBAN	
AM 640	WMFN	IIIII	SPRTS	
AM 1230	WTKG	I	NWS, TLK	
AM 1300	WOOD	IIIII	NWS, TLK	
AM 1340	WBBL	I	SPRTS	

KALAMAZOO, MI

FREQ	STATION	POWER	FORMAT	
103.3	WKFR	IIIII	AC	
106.5	WQLR	IIIII	AC	
96.5	WFAT	I	TOP 40	
107.7	WRKR	IIIII	CLS RK	
89.1	WIDR	I	AC	C
102.1	WMUK	IIIII	CLS, NPR, PRI	P
AM 1420	WKPR	I	RLG	
AM 590	WKZO	IIIIIIII	NWS, TLK	
AM 1360	WKMI	IIIII	NWS, TLK	
AM 1470	WQSN	I	SPRTS	

LANSING, MI

FREQ	STATION	POWER	FORMAT	
99.1	WFMK	IIIIIIII	AC	
101.7	WHZZ	III	TOP 40	
92.1	WWDX	III	DVRS	
94.9	WMMQ	IIIII	CLS RK	
97.5	WJIM	IIIII	OLDS	
100.7	WITL	IIIII	CTRY	
AM 1320	WILS	IIIII	BB, NOS	
88.9	WDBM	III	ALT RK	C
89.7	WLNZ	I	JZZ, CLS, PRI	P
90.5	WKAR	IIIIIIII	CLS, NPR, PRI	P
AM 870	WKAR	IIIIIIII	NWS, TLK, NPR	P
96.5	WQHH	III	URBAN	
AM 1240	WJIM	I	NWS, TLK	

NORTHWEST MICH.

FREQ	STATION	POWER	FORMAT	
96.3	WLXT	IIIIIIII	LGHT AC	
99.3	WBNZ	IIIII	HOT AC	
101.9	WLDR	IIIIIIII	AC	
105.9	WKHQ	IIIIIIII	HOT AC	
95.5	WJZJ	IIIIIIII	MOD RK	
97.5	WKLT	IIIII	ROCK	
98.9	WKLZ	IIIIIIII	ROCK	
107.5	WCCW	IIIIIIII	OLDS	

NORTHWEST MICH • SAGINAW, MI • MORRISTOWN
ATLANTIC CITY • MONMTH/OCEAN CTY • TRENTON, NJ

	FREQ	STATION	POWER	FORMAT	
AM	1340	WMBN	I	OLDS	
	93.5	WBCM	III	CTRY	
	94.3	WBYB	IIIIIIII	CTRY	
	103.5	WTCM	IIIIIIII	CTRY	
	88.7	WIAA	IIIIIIII	CLS, NPR, PRI	P
	90.7	WNMC	III	DVRS	C
	100.9	WIZY	III	CLS, NPR, PRI	P
	103.9	WCMW	IIIII	CLS, NPR, PRI	P
AM	580	WTCM		NWS, TLK	

SAGINAW, MI

	102.5	WIOG	IIIIIIII	AC	
	106.3	WGER	III	AC	
	100.5	WTCF	III	TOP 40	
	93.3	WKQZ	IIIII	ROCK	
	97.3	WIXC	III	CLS RK	
	100.9	WMJK	III	70's RK	
	104.5	WMJA	III	70's RK	
	96.1	WHNN	IIIIIIII	OLDS	
	98.1	WKCQ	IIIII	CTRY	
AM	1400	WSAM	I	BB, NOS	
AM	1490	WMPX	I	BB, NOS	
	90.1	WUCX	III	CLS, NPR, PRI	P
	107.1	WTLZ	III	URBAN	
AM	790	WSGW	IIIIIIII	NWS, TLK	
AM	1440	WMAX	IIIII	SPRTS	

ATLANTIC CITY, NJ

	95.1	WAYV	IIIII	AC	
	96.1	WTTH	III	AC	
	96.9	WFPG	IIIII	LGHT AC	
	103.7	WMGM	IIIII	CLS RK	
	99.3	WSAX	III	JZZ	
	104.9	WRDR	III	BB, NOS	
AM	1340	WMID	I	BB, NOS	
	89.7	WNJN	III	NWS, TLK, NPR, PRI	P
	91.7	WLFR	I	DVRS	C
AM	1400	WOND	I	NWS, TLK	
AM	1450	WFPG	I	NWS, TLK	

MON/OCEAN CTY, NJ

	92.7	WOBM	III	AC	
	94.3	WJLK	III	HOT AC	
	100.1	WJRZ	III	HOT AC	
	98.5	WBBO	III	TOP 40	
	95.9	WRAT	III	CLS RK	
	106.3	WHTG	III	MOD RK	
	107.1	WWZY	III	CTRY	
AM	1160	WOBM	IIIII	BB, NOS	
AM	1310	WADB	III	BB, NOS	
	89.3	WCNJ	III	AC	P
	90.5	WBJB	III	ROCK	C

MORRISTOWN, NJ

	105.5	WDHA	III	ROCK	
AM	1250	WMTR	IIIII	BB, NOS	
	91.1	WWNJ	III	CLS	P

TRENTON, NJ

	97.5	WPST	IIIII	TOP 40	
	94.5	WNJO	IIIII	OLDS	
	101.5	WKXW	IIIIIIII	OLDS	

MI NJ

27

	FREQ	STATION	POWER	FORMAT	
AM	1260	WBUD	IIIII	OLDS	
	88.1	WNJT	III	NWS, TLK	P
	89.1	WWFM	I	CLS, NPR, PRI	P
	91.3	WTSR	I	DVRS	C
AM	1300	WIMG	IIIII	GOSP	
AM	920	WCHR	III	RLG	
AM	1350	WHWH	IIIII	NWS, TLK	

AKRON, OH

	FREQ	STATION	POWER	FORMAT	
	96.5	WKDD	IIIII	HOT AC	
	97.5	WONE	III	CLS RK	
AM	1590	WAKR	IIIII	OLDS	
	89.7	WKSU	IIIII	CLS, NPR, PRI	P
AM	640	WHLO	IIIIIIII	RLG	
	88.1	WZIP	III	URBAN	P
	100.1	WNIR	III	NWS, TLK	
AM	1350	WTOU	IIIII	URBAN	

CANTON, OH

	FREQ	STATION	POWER	FORMAT	
AM	1480	WHBC	IIIII	AC	
	106.9	WRQK	III	MOD RK	
	92.5	WZKL	IIIII	OLDS	
	94.1	WHBC	IIIIIIII	BB, NOS	
AM	1310	WDPN	I	BB, NOS	
	91.1	WRMU	I	JZZ	C
	98.1	WHK	IIIII	RLG	
AM	990	WTIG	III	SPRTS	

CINCINNATI, OH

	FREQ	STATION	POWER	FORMAT	
	94.1	WVMX	IIIII	AC	
	98.5	WRRM	III	AC	
	101.9	WKRQ	III	TOP 40	
	107.1	WKFS	III	TOP 40	
	92.5	WOFX	IIIIIIII	CLS RK	
	97.3	WYLX	III	CLS RK	
	97.7	WOXY	III	MOD RK	
	102.7	WEBN	IIIIIIII	ROCK	
	103.5	WGRR	III	OLDS	
	96.5	WYGY	IIIIIIII	CTRY	
	99.3	WSCH	III	CTRY	
	105.1	WUBE	IIIIIIII	CTRY	
	106.5	WNKR	III	CTRY	
AM	1480	WCIN	IIIII	BLK, R&B	
	94.9	WVAE	III	JZZ	
AM	1530	WSAI	IIIIIIII	BB, NOS	
	88.5	WMUB	III	BB, NOS, NPR, PRI	C
	88.7	WOBO	III	DVRS	P
	89.7	WNKU	III	DVRS, NPR, PRI	P
	90.9	WGUC	III	CLS, NPR, PRI	P
	91.7	WVXU	IIIII	DVRS, NPR, PRI	P
	104.3	WNLT	IIIII	RLG AC	
AM	1320	WCVG	I	GOSP	
	107.5	WIOK	III	GOSP	
	100.9	WIZF	III	URBAN	
AM	550	WKRC	IIIIIIII	NWS, TLK	
AM	700	WLW	IIIIIIII	NWS, TLK	
AM	1160	WBOB	IIIII	SPRTS	
AM	1230	WUBE	I	SPRTS	
AM	1360	WCKY	IIIII	SPRTS	
AM	1450	WMOH	I	NWS, TLK	

CLEVELAND, OH

	FREQ	STATION	POWER	FORMAT	
	102.1	WDOK	IIIII	AC	
	104.1	WQAL	IIIII	HOT AC	

28

OH

	FREQ	STATION	POWER	FORMAT	
	106.5	WMVX	III	AC	
	92.3	WZJM	IIIII	TOP 40	
	98.5	WNCX	III	CLS RK	
	100.7	WMMS	IIIIIIII	ROCK	
	107.9	WENZ	IIIIIIII	MOD RK	
	105.7	WMJI	IIIII	OLDS	
	94.9	WQMX	III	CTRY	
	99.5	WGAR	IIIII	CTRY	
AM	1320	WOBL	I	CTRY	
AM	1490	WJMO	IIIIIIII	BLK, R&B	
	95.5	WCLV	IIIII	CLS	
	107.3	WNWV	IIIII	JZZ	
AM	850	WRMR	IIIIIIII	BB, NOS	
	88.3	WBWC	I	AC	C
	88.7	WJCU	III	DVRS	C
	89.1	WKSV	IIIII	NWS, TLK, PRI	P
	89.3	WCSB	I	DVRS	C
	90.3	WCPN	IIIII	JZZ, NPR, PRI	P
	91.1	WRUW	III	DVRS	C
	91.5	WOBC	I	DVRS	C
	104.9	WZLE	III	RLG AC	
AM	1330	WELW	I	DVRS	
AM	1260	WMIH	IIIII	KIDS	
AM	1420	WHK	IIIII	RLG	
	93.1	WZAK	IIIII	URBAN	
AM	930	WEOL	III	NWS, TLK	
AM	1100	WTAM	IIIIIIII	NWS, TLK	
AM	1220	WKNR	IIIIIIII	SPRTS	
AM	1300	WERE	IIIII	NWS, TLK	

COLUMBUS, OH

	FREQ	STATION	POWER	FORMAT	
	94.7	WSNY	III	AC	
	107.1	WAZU	III	HOT AC	
	97.9	WNCI	IIIIIIII	TOP 40	
	96.3	WLVQ	III	ROCK	
	99.7	WBZX	III	MOD RK	
	101.1	WWCD	III	DVRS	
	105.7	WZAZ	III	DVRS	
	107.9	WXST	III	CLS RK	
	97.1	WBNS	IIIIIIII	OLDS	
	101.7	WNKO	III	OLDS	
	92.3	WCOL	IIIIIIII	CTRY	
	95.5	WHOK	IIIII	CTRY	
	100.3	WCLT	IIIII	CTRY	
	103.5	WJZA	III	JZZ	
	104.3	WZJZ	I	JZZ	
AM	920	WMNI	III	BB, NOS	
	89.7	WOSU	III	CLS, NPR, PRI	P
	91.1	WDUB	I	DVRS	C
	98.7	WSLN	I	DVRS	C
AM	820	WOSU	IIIIIIII	NWS, TLK, NPR, PRI	P
AM	880	WRFD	IIIIIIII	RLG AC	
AM	1580	WVKO	I	GOSP	
	98.9	WMXG	III	URBAN	
	103.1	WSMZ	III	URBAN	
	105.7	WXMG	III	URBAN	
	107.5	WCKX	III	URBAN	
AM	610	WTVN	IIIIIIII	NWS, TLK	
AM	1230	WFII	I	NWS, TLK	
AM	1460	WBNS	IIIII	SPRTS	

DAYTON, OH

	FREQ	STATION	POWER	FORMAT	
	99.9	WLQT	IIIII	AC	
	107.7	WMMX	IIIII	HOT AC	

FREQ	STATION	POWER	FORMAT	
92.9	WGTZ	IIIII	TOP 40	
94.5	WBTT	III	TOP 40	
102.9	WING	IIIII	CLS RK	
104.7	WTUE	IIIII	CLS RK	
95.3	WZLR	III	OLDS	
95.7	WCLR	IIIII	OLDS	
96.9	WRNB	III	OLDS	
99.1	WHKO	IIIIIIII	CTRY	
AM 980	WONE	IIIIIIII	BB, NOS	
AM 1340	WIZE	I	BB, NOS	
88.1	WDPR	I	CLS, NPR, PRI	P
88.9	WCSU	I	URBAN	C
91.3	WYSO	III	DVRS, NPR, PRI	C
93.7	WFCJ	IIIII	RLG	
92.1	WROU	III	URBAN	
AM 1290	WHIO	IIIII	NWS, TLK	
AM 1410	WING	IIIII	NWS, TLK	

LIMA, OH

FREQ	STATION	POWER	FORMAT	
104.9	WAJC	III	AC	
92.1	WZOQ	III	TOP 40	
107.5	WBUK	III	OLDS	
93.1	WFGF	III	CTRY	
102.1	WIMT	IIIIIIII	CTRY	
107.1	WDOH	III	CTRY	
90.7	WGLE	IIIII	NWS, NPR, PRI	P
AM 1150	WIMA	I	NWS, TLK	

TOLEDO, OH

FREQ	STATION	POWER	FORMAT	
96.1	WMTR	III	AC	
101.5	WRVF	III	AC	
105.5	WWWM	III	AC	
92.5	WVKS	IIIII	TOP 40	
103.9	WXEG	III	DVRS	
104.7	WIOT	IIIIIIII	ROCK	
106.5	WBUZ	III	ROCK	
93.5	WRQN	III	OLDS	
99.9	WKKO	IIIII	CTRY	
107.7	WHMQ	III	CTRY	
AM 1230	WCWA	I	BB, NOS	
88.1	WBGU	I	DVRS	C
88.3	WXUT	I	DVRS	C
91.3	WGTE	III	NWS, NPR, PRI	P
AM 1520	WDMN	I	GOSP	
107.3	WJUC	III	URBAN	
AM 1370	WSPD	IIIII	NWS, TLK	
AM 1470	WLQR	I	SPRTS	

YOUNGSTOWN, OH

FREQ	STATION	POWER	FORMAT	
98.9	WKBN	III	HOT AC	
101.1	WHOT	IIIIIIII	TOP 40	
106.1	WNCD	III	CLS RK	
93.3	WBBG	IIIII	OLDS	
88.5	WYSU	IIIII	CLS, NPR, PRI	P
101.9	WBTJ	III	URBAN	
AM 570	WKBN	IIIIIIII	NWS, TLK	
AM 1240	WBBW	I	SPRTS	
AM 1390	WRTK	IIIII	NWS, TLK	
AM 1440	WRRO	IIIII	SPRTS	

ALLENTOWN, PA

FREQ	STATION	POWER	FORMAT	
100.7	WLEV	IIIII	AC	
104.1	WAEB	IIIII	TOP 40	
95.1	WZZO	IIIII	ROCK	

	FREQ	STATION	POWER	FORMAT	
	99.9	WODE	IIIII	OLDS	
AM	1470	WKAP	IIIII	OLDS	
AM	1510	WRNJ	III	OLDS	
	96.1	WCTO	IIIII	CTRY	
	107.1	WWYY	III	CTRY	
AM	1160	WYNS	IIIII	CTRY	
AM	1400	WEST	I	BB, NOS	
AM	1410	WLSH	IIIII	BB, NOS	
	88.1	WDIY	I	NWS, NPR	P
	90.3	WXLV	I	DVRS	C
	91.3	WLVR	I	DVRS	C
	91.7	WMUH	I	DVRS	C
	91.9	WNTI	III	DVRS	C
AM	1600	WHOL	I	RLG AC	
AM	790	WAEB	IIIII	NWS, TLK	
AM	1230	WEEX	I	NWS, TLK	
AM	1320	WTKZ	IIIII	SPRTS	

ALTOONA, PA

	FREQ	STATION	POWER	FORMAT	
	104.9	WMXV	III	AC	
	100.1	WPRR	IIIIIIII	TOP 40	
	101.1	WGMR	IIIIIIII	MOD RK	
	103.9	WALY	III	OLDS	
	98.1	WFGY	IIIIIIII	CTRY	
AM	1290	WFBG	IIIII	BB, NOS	
AM	1340	WTRN	I	EZ	
AM	1240	WRTA	I	NWS, TLK	
AM	1430	WVAM	IIIII	SPRTS	

ERIE, PA

	FREQ	STATION	POWER	FORMAT	
	99.9	WXKC	IIIII	AC	
	103.7	WRTS	IIIII	AC	
	102.3	WJET	III	TOP 40	
	100.9	WRKT	III	CLS RK	
	94.7	WFGO	III	OLDS	
	97.9	WXTA	IIIII	CTRY	
AM	1260	WRIE	IIIII	BB, NOS	
	88.5	WMCE	III	DVRS	C
	88.9	WFSE	III	DVRS	C
	89.9	WERG	I	ROCK	C
	91.3	WQLN	III	CLS, NPR, PRI	P
	106.3	WCTL	III	RLG AC	
AM	1330	WFLP	IIIII	NWS, TLK	
AM	1400	WLKK	I	NWS, TLK	
AM	1450	WPSE	I	BUS NWS	

HARRISBURG, PA

	FREQ	STATION	POWER	FORMAT	
	97.3	WRVV	III	AC	
	98.9	WQLV	III	AC	
	100.1	WQIC	III	LGHT AC	
AM	1270	WLBR	IIIII	AC	
	104.1	WNNK	III	TOP 40	
	93.5	WTPA	III	ROCK	
	99.3	WWKL	IIIIIIII	OLDS	
AM	1460	WWKL	IIIII	OLDS	
	94.9	WRBT	III	CTRY	
	100.5	WYGL	III	CTRY	
	102.3	WHYL	III	CTRY	
	106.7	WRKZ	IIIIIIII	CTRY	
AM	960	WHYL	IIIIIIII	BB, NOS	
AM	1230	WKBO	I	BB, NOS	
	88.1	WXPH	I	DVRS, PRI	C

PA

	FREQ	STATION	POWER	FORMAT	
	88.3	WDCV	I	DVRS	C
	88.7	WSYC	I	DVRS	C
	89.5	WITF	III	CLS, NPR, PRI	P
	89.7	WQEJ	I	CLS	P
	91.7	WJAZ	III	JZZ, NPR	P
	90.7	WVMM	I	RLG AC	C
	92.1	WNCE	III	EZ	
AM	580	WHP	IIIIIIII	NWS, TLK	
AM	1400	WTCY	I	URBAN	

JOHNSTOWN, PA

	FREQ	STATION	POWER	FORMAT	
	95.5	WKYE	IIIIIIII	AC	
	101.7	WSRA	III	LGHT AC	
	92.1	WGLU	III	TOP 40	
	99.1	WQKK	IIIII	DVRS	
AM	850	WJAC	IIIIIIII	OLDS	
	96.5	WMTZ	IIIII	CTRY	
	97.7	WSGY	III	CTRY	
	105.7	WFJY	III	CTRY	
AM	1490	WNTJ	I	NWS, TLK	

LANCASTER, PA

	FREQ	STATION	POWER	FORMAT	
	101.3	WROZ	IIIIIIII	AC	
	96.9	WLAN	IIIII	TOP 40	
	105.1	WIOV	IIIII	CTRY	
	91.3	WLCH	I	SPAN	P
AM	1390	WLAN	IIIII	BB, NOS	
	88.3	WWEC	III	DVRS	C
	89.1	WFNM	I	DVRS	C
	91.7	WIXQ	I	AC	C
	94.5	WDAC	IIIIIIII	RLG	
AM	1490	WLPA	I	SPRTS	

PHILADELPHIA, PA

	FREQ	STATION	POWER	FORMAT	
	95.7	WXXM	IIIIIIII	AC	
	101.1	WBEB	IIIIIIII	AC	
	104.5	WYXR	III	HOT AC	
	102.1	WIOQ	IIIII	TOP 40	
	93.3	WMMR	IIIII	ROCK	
	94.1	WYSP	III	MOD RK	
	100.3	WPLY	IIIII	MOD RK	
	102.9	WMGK	III	CLS RK	
	98.1	WOGL	IIIII	OLDS	
	92.5	WXTU	III	CTRY	
AM	860	WTEL	IIIIIIII	SPAN	
	106.1	WJJZ	III	JZZ	
AM	950	WPEN	IIIIIIII	BB, NOS	
	88.1	WNJS	III	NWS, TLK, NPR, PRI	P
	88.5	WXPN	III	DVRS, PRI	C
	88.9	WBZC	I	DVRS	C
	89.1	WYBF	I	DVRS	C
	89.7	WGLS	I	DVRS	C
	90.1	WRTI	IIIII	JZZ, NPR	P
	90.9	WHYY	III	NWS, TLK, NPR, PRI	P
	91.1	WRTY	I	JZZ, NPR	P
	91.5	WSRN	I	AC	C
	91.5	WDBK	I	DVRS	C
	91.7	WKDU	I	AC	C
AM	990	WZZD	IIIIIIII	RLG AC	
AM	1480	WDAS	IIIII	RLG AC	
AM	560	WFIL	IIIIIIII	RLG	
AM	1540	WNWR	IIIIIIII	ETH	

	FREQ	STATION	POWER	FORMAT	
	96.5	WWDB	IIIIIIII	NWS, TLK	
	98.9	WUSL	III	URBAN	
	103.9	WPHI	III	URBAN	
	105.3	WDAS	III	URBAN	
AM	610	WIP	IIIIIIII	NWS, TLK	
AM	1060	KYW	IIIIIIII	NWS	
AM	1210	WPHT	IIIIIIII	NWS, TLK	
AM	1340	WHAT	I	NWS, TLK	

PITTSBURGH, PA

	FREQ	STATION	POWER	FORMAT	
	92.9	WLTJ	IIIIIIII	LGHT AC	
	96.1	WDRV	IIIII	HOT AC	
	99.3	WPQR	III	AC	
	99.7	WSHH	III	AC	
	100.7	WZPT	III	AC	
	103.9	WLSW	III	AC	
AM	590	WMBS	IIIII	AC	
AM	1340	WCVI	I	AC	
	93.7	WBZZ	IIIII	TOP 40	
	98.3	WZKT	I	TOP 40	
	96.9	WRRK	IIIII	CLS RK	
	102.5	WDVE	IIIIIIII	CLS RK	
	105.9	WXDX	IIIII	DVRS	
	94.5	WWSW	IIIIIIII	OLDS	
	95.3	WJPA	III	OLDS	
AM	970	WWSW	IIIIIIII	OLDS	
AM	1450	WJPA	I	OLDS	
	94.9	WASP	IIIIIIII	CTRY	
	107.9	WDSY	IIIII	CTRY	
	104.7	WJJJ	IIIIIIII	JZZ	
AM	1320	WJAS	IIIII	BB, NOS	
	88.1	WRSK	I	ROCK	C
	88.3	WRCT	I	DVRS	C
	89.3	WQED	III	CLS, NPR, PRI	P
	90.5	WDUQ	III	JZZ, NPR, PRI	P
	91.3	WYEP	III	ALT RK, NPR, PRI	P
	91.9	WVCS	I	AC	C
	101.5	WORD	IIIII	RLG AC	
	90.1	WSRU	III	EDU	C
	106.7	WAMO	III	URBAN	
	107.1	WSSZ	III	URBAN	
AM	860	WAMO	III	URBAN	
AM	1020	KDKA	IIIIIIII	NWS, TLK	
AM	1230	WBVP	I	NWS, TLK	
AM	1250	WEAE	IIIII	SPRTS	
AM	1360	WPTT	IIIII	NWS, TLK	
AM	1410	KQV	IIIII	NWS	

READING, PA

	FREQ	STATION	POWER	FORMAT	
AM	850	WEEU	III	AC	
	102.5	WRFY	III	TOP 40	
AM	1340	WRAW	I	BB, NOS	
	91.3	WXAC	I	ROCK	C
	107.5	WBYN	IIIIIIII	RLG AC	
AM	1240	WIOV	I	NWS, TLK	

SCRANTON, PA

	FREQ	STATION	POWER	FORMAT	
	92.9	WMGS	IIIIIIII	AC	
	93.5	WSBG	III	HOT AC	
	104.9	WWDL	III	AC	
	106.5	WHLM	IIIII	AC	
	97.1	WBHT	III	TOP 40	
	98.5	WKRZ	III	TOP 40	
	107.7	WEMR	III	TOP 40	

PA

	FREQ	STATION	POWER	FORMAT	
	107.9	WKRF	III	TOP 40	
	95.7	WKQV	III	ROCK	
	97.9	WZMT	IIIIIIII	ROCK	
	106.9	WEZX	III	CLS RK	
	92.1	WQFM	III	OLDS	
	103.5	WKAB	III	OLDS	
AM	1340	WYCK	I	OLDS	
AM	1400	WICK	I	OLDS	
	93.7	WCTP	III	CTRY	
	94.3	WCTD	III	CTRY	
	101.3	WGGY	III	CTRY	
AM	550	WJMW	IIIII	BB, NOS	
AM	1240	WBAX	I	BB, NOS	
AM	1490	WAZL	I	BB, NOS	
	88.5	WRKC	I	ROCK	C
	89.9	WVIA	III	CLS, NPR, PRI	P
	90.3	WESS	I	DVRS, PRI	C
	90.7	WCLH	III	AC	C
	91.1	WBUQ	I	AC	C
	91.5	WVMW	I	AC	C
	99.5	WUSR	I	DVRS	C
	102.3	WILK	I	NWS, TLK	
	103.1	WILP	III	NWS, TLK	
AM	590	WARM	IIIIIIII	NWS, TLK	
AM	910	WGBI	III	NWS, TLK	
AM	980	WILK	IIIIIIII	NWS, TLK	

STATE COLLEGE, PA

	FREQ	STATION	POWER	FORMAT	
	95.3	WZWW	III	AC	
	105.9	WUBZ	I	AC	
	107.9	WIKN	III	HOT AC	
	103.1	WBHV	I	TOP 40	
	97.1	WQWK	III	DVRS	
	90.7	WKPS	I	DVRS	C
	91.5	WPSU	III	CLS, NPR, PRI	C
AM	1390	WRSC	III	NWS, TLK	
AM	1450	WMAJ	I	NWS, TLK	

WILLIAMSPORT, PA

	FREQ	STATION	POWER	FORMAT	
	97.7	WVRT	III	HOT AC	
	102.7	WKSB	IIIIIIII	AC	
	107.9	WSFT	III	LGHT AC	
	93.3	WHTO	III	TOP 40	
	99.3	WZXR	III	CLS RK	
AM	1340	WWPA	I	OLDS	
	105.1	WILQ	III	CTRY	
	95.5	WMYL	I	BB, NOS	
	88.1	WPTC	I	DVRS	C
	91.7	WRLC	I	AC	C
	96.3	WJSA	I	RLG AC	
AM	1400	WRAK	I	NWS, TLK	

YORK, PA

	FREQ	STATION	POWER	FORMAT	
	103.3	WARM	III	AC	
AM	1320	WGET	I	AC	
	98.5	WYCR	IIIIIIII	TOP 40	
	92.7	WEGK	III	CLS RK	
	105.7	WQXA	IIIII	DVRS	
	96.1	WSOX	IIIII	OLDS	
	107.7	WGTY	IIIIIIII	CTRY	
AM	1350	WOYK	IIIII	CTRY	
AM	1250	WQXA	I	BB, NOS	
AM	1280	WHVR	IIIII	BB, NOS	
	88.1	WVYC	I	AC	C

	FREQ	STATION	POWER	FORMAT	
	91.1	WZBT	I	DVRS	C
AM	910	WSBA	IIIIIIII	NWS, TLK	

CHARLOTTESVILLE, VA

	FREQ	STATION	POWER	FORMAT	
	92.7	WUVA	III	AC	
	107.5	WUMX	III	AC	
	95.1	WQMZ	I	TOP 40	
	97.5	WWWV	IIIIIIII	CLS RK	
	102.3	WVAO	III	OLDS	
	88.5	WVTW	III	DVRS, NPR	P
	89.3	WVTU	III	CLS, NPR, PRI	P
	91.1	WTJU	III	DVRS	C
	91.9	WNRN	III	MOD RK	P
	103.5	WMRY	I	CLS, NPR, PRI	P
AM	1400	WKAV	I	DVRS	
	101.9	WVSY	III	EZ	
AM	1070	WINA	IIIII	NWS, TLK	
AM	1260	WCHV	IIIII	URBAN	

HARRISONBURG, VA

	FREQ	STATION	POWER	FORMAT	
	100.7	WQPO	IIIII	TOP 40	
	98.5	WACL	IIIIIIII	CLS RK	
	104.3	WKCY	IIIII	CTRY	
	105.1	WAMM	III	CTRY	
	90.7	WMRA	IIIII	CLS, NPR, PRI	P
	96.1	WLTK	III	RLG AC	
	91.7	WEMC	I	RLG	C
AM	550	WSVA	IIIIIIII	NWS, TLK	
AM	1360	WHBG	IIIII	NWS, TLK	

NORFOLK, VA

	FREQ	STATION	POWER	FORMAT	
	92.9	WFOG	IIIII	AC	
	94.9	WPTE	IIIII	AC	
	99.1	WXGM	III	AC	
	101.3	WWDE	IIIII	AC	
	104.5	WNVZ	IIIII	TOP 40	
AM	1270	WTJZ	I	TOP 40	
	98.7	WNOR	IIIIIIII	ROCK	
	106.9	WAFX	IIIIIIII	CLS RK	
AM	1230	WNOR	IIIII	ROCK	
	95.7	WVKL	IIIII	OLDS	
	97.3	WGH	IIIII	CTRY	
	100.5	WCMS	IIIII	CTRY	
AM	1050	WCMS	IIIII	CTRY	
	105.3	WJCD	IIIII	JZZ	
	88.1	WHOV	I	JZZ	C
	89.5	WHRV	III	NWS, NPR, PRI	P
	90.3	WHRO	III	CLS, NPR, PRI	P
	90.7	WCWM	III	JZZ	C
	91.1	WNSB	III	JZZ, NPR	C
AM	1350	WGPL	IIIII	GOSP	
AM	1400	WPCE	I	GOSP	
	94.1	WXEZ	IIIII	EZ	
AM	1010	WPMH	IIIII	RLG	
	92.1	WSVV	III	URBAN	
	102.9	WOWI	IIIIIIII	URBAN	
AM	790	WNIS	IIIIIIII	NWS, TLK	
AM	850	WTAR	IIIIIIII	SPRTS	
AM	1310	WGH	IIIII	SPRTS	

RICHMOND, VA

	FREQ	STATION	POWER	FORMAT	
	98.1	WTVR	IIIIIIII	LGHT AC	

PA
VA

35

VA
WI

	FREQ	STATION	POWER	FORMAT	
	103.7	WMXB	III	HOT AC	
	94.5	WRVQ	IIIII	TOP 40	
	102.1	WRXL	IIIII	CLS RK	
	106.5	WBZU	III	DVRS	
AM	1450	WCLM	I	OLDS	
	95.3	WKHK	III	CTRY	
AM	950	WXGI	IIIIIIII	CTRY	
	101.1	WJRV	I	JZZ	
AM	1380	WTVR	IIIII	BB, NOS	
	88.9	WCVE	III	CLS, NPR, PRI	P
	90.1	WDCE	I	DVRS	C
	91.3	WVST	I	JZZ	C
	105.7	WDYL	III	RLG AC	
AM	820	WGGM	IIIIIIII	RLG	
AM	1240	WGCV	I	RLG	
	92.1	WCDX	III	URBAN	
	99.3	WPLZ	III	URBAN	
	100.3	WSOJ	III	URBAN	
AM	910	WRNL	IIIIIIII	SPRTS	
AM	1140	WRVA	IIIIIIII	NWS, TLK	

ROANOKE, VA

	FREQ	STATION	POWER	FORMAT	
	97.9	WRVX	III	AC	
	99.1	WSLQ	IIIIIIII	AC	
	92.3	WXLK	IIIIIIII	TOP 40	
	100.1	WLYK	III	TOP 40	
	101.7	WJJX	III	TOP 40	
	106.1	WJJS	III	TOP 40	
AM	1240	WGMN	I	ROCK	
	94.9	WPVR	IIIIIIII	OLDS	
	102.7	WLDJ	IIIIIIII	OLDS	
	104.9	WRDJ	III	OLDS	
AM	1350	WBLT	I	OLDS	
	93.5	WJLM	III	CTRY	
	101.5	WZZI	III	CTRY	
	105.5	WKDE	III	CTRY	
	107.9	WYYD	IIIII	CTRY	
AM	610	WSLC	IIIIIIII	CTRY	
	106.9	WLQE	III	BB, NOS	
AM	590	WLVA	IIIIIIII	BB, NOS	
AM	880	WVLR	III	BB, NOS	
	89.1	WVTF	IIIIIIII	CLS, NPR, PRI	P
	89.9	WMRL	I	CLS, NPR, PRI	P
	105.9	WLNI	III	NWS, TLK	
AM	960	WFIR	IIIIIIII	NWS, TLK	
AM	1480	WTOY	IIIII	URBAN	

APPLETON / OSHKOSH, WI

	FREQ	STATION	POWER	FORMAT	
	94.3	WROE	III	AC	
	93.5	WOZZ	IIIII	CLS RK	
	105.7	WAPL	IIIIIIII	ROCK	
	103.1	WOGB	III	OLDS	
	103.9	WVBO	III	OLDS	
	96.9	WUSW	III	CTRY	
	100.3	WNCY	IIIII	CTRY	
	104.9	WPCK	III	CTRY	
AM	1280	WNAM	IIIII	BB, NOS	
AM	1570	WRJQ	I	BB, NOS	
	90.3	WRST	I	DVRS, NPR, PRI	C
	91.1	WLFM	III	NWS, TLK, NPR, PRI	P
AM	1150	WHBY	IIIII	NWS, TLK	
AM	1490	WOSH	I	NWS, TLK	

36

WI

	FREQ	STATION	POWER	FORMAT	
	105.7	WCFW	III	AC	
	94.1	WIAL	IIIII	TOP 40	
	100.7	WBIZ	IIIII	TOP 40	
	98.1	WISM	IIIII	CLS RK	
	95.1	WQRB	III	CTRY	
	104.5	WAXX	IIIIIIII	CTRY	
	89.7	WUEC	III	CLS, NPR, PRI	C
AM	680	WOGO	IIIIIIII	RLG AC	
AM	790	WAYY	IIIIIIII	NWS, TLK	
AM	1400	WBIZ	I	SPRTS	

GREEN BAY, WI

	FREQ	STATION	POWER	FORMAT	
	95.9	WKSZ	III	TOP 40	
	98.5	WQLH	IIIII	TOP 40	
	101.1	WIXX	IIIIIIII	TOP 40	
	106.7	WJLW	III	CLS RK	
	88.1	WHID	IIIII	NWS, TLK, NPR, PRI	P
	89.3	WPNE	III	CLS, NPR, PRI	P
AM	1360	WGEE	IIIII	NWS, TLK	
AM	1400	WDUZ	I	SPRTS	
AM	1440	WNFL	IIIII	NWS, TLK	

LA CROSSE, WI

	FREQ	STATION	POWER	FORMAT	
	104.9	WLXR	III	AC	
	93.3	WIZM	IIIIIIII	TOP 40	
	95.7	WRQT	IIIII	CLS RK	
	100.1	WKBH	III	CLS RK	
	106.3	WQCC	III	CTRY	
AM	1490	WLFN	I	BB, NOS	
	88.9	WLSU	III	CLS, NPR, PRI	P
	90.3	WHLA	III	NWS, TLK, NPR, PRI	P
AM	580	WKTY	IIIIIIII	SPRTS	
AM	1410	WIZM	IIIII	NWS, TLK	
AM	1570	WKBH	I	SPRTS	

MADISON, WI

	FREQ	STATION	POWER	FORMAT	
	98.1	WMGN	IIIIIIII	AC	
	104.1	WZEE	IIIII	TOP 40	
	92.1	WMAD	III	DVRS	
	101.5	WIBA	IIIIIIII	ROCK	
	105.5	WMMM	III	ALT RK	
	105.1	WYZM	III	CTRY	
	106.3	WWQM	III	CTRY	
AM	1070	WTSO	IIIII	BB, NOS	
	88.7	WERN	IIIII	NWS, NPR, PRI	P
	89.9	WORT	III	DVRS, PRI	P
AM	970	WHA	IIIIIIII	NWS, TLK, NPR	P
	102.5	WNWC	IIIII	RLG	C
AM	1190	WNWC	I	RLG	C
AM	1310	WIBA	IIIII	NWS, TLK	
AM	1480	WTDY	IIIII	NWS, TLK	

MILWAUKEE, WI

	FREQ	STATION	POWER	FORMAT	
	94.5	WKTI	IIIIIIII	HOT AC	
	97.3	WLTQ	III	AC	
	99.1	WMYX	IIIII	HOT AC	
	106.9	WPNT	I	AC	
AM	1290	WMCS	IIIII	AC	
	103.7	WXSS	IIIIIIII	TOP 40	
	96.5	WKLH	IIIIIIII	CLS RK	

	FREQ	STATION	POWER	FORMAT	
	102.1	WLUM	|||||||	MOD RK	
	102.9	WLZR	|||||	ROCK	
	95.7	WZTR	|||||	OLDS	
AM	1250	WEMP	|||||	OLDS	
	92.5	WBWI	III	CTRY	
	104.7	WEXT	III	CTRY	
	104.9	WTKM	III	CTRY	
	106.1	WMIL	|||||||	CTRY	
	93.3	WJZI	III	JZZ	
	98.3	WFMR	III	CLS	
AM	920	WOKY	|||||||	BB, NOS	
	89.7	WUWM	III	NWS, NPR, PRI	P
	90.7	WHAD	|||||||	NWS, TLK, NPR, PRI	P
	91.7	WMSE	I	DVRS	C
AM	1340	WJYI	I	RLG AC	
	92.1	WEZY	III	EZ	
AM	1400	WRJN	I	FS	
	100.7	WKKV	|||||	URBAN	
AM	620	WTMJ	|||||||	NWS, TLK	
AM	1130	WISN	|||||||	NWS, TLK	
AM	1470	WBKV	III	NWS, TLK	
AM	1510	WAUK	|||||	SPRTS	

BECKLEY, VA

	FREQ	STATION	POWER	FORMAT	
	97.9	WSPT	|||||	HOT AC	
	106.5	WLJY	|||||||	LGHT AC	
	107.9	WYCO	|||||||	HOT AC	
	95.5	WIFC	|||||||	TOP 40	
	103.3	WGLX	|||||	CLS RK	
	104.9	WKQH	III	CLS RK	
	94.7	WOFM	|||||||	OLDS	
	99.9	WIZD	III	OLDS	
	92.3	WOSQ	III	CTRY	
	96.7	WYTE	|||||||	CTRY	
	101.9	WDEZ	|||||	CTRY	
AM	1390	WRIG	|||||	BB, NOS	
	89.9	WWSP	III	DVRS	C
	90.9	WHRM	|||||||	CLS, NPR, PRI	P
	91.9	WXPW	III	DVRS, NPR, PRI	P
AM	930	WLBL	|||||||	NWS, TLK, NPR, PRI	P
AM	550	WSAU	|||||||	NWS, TLK	
AM	1230	WXCO	I	NWS, TLK	
AM	1320	WFHR	|||||	NWS, TLK	
AM	1450	WDLB	I	NWS, TLK	

BECKLEY, WV

	FREQ	STATION	POWER	FORMAT	
	103.7	WCIR	III	AC	
AM	1070	WIWS	|||||	OLDS	
	99.5	WJLS	|||||||	CTRY	
AM	560	WJLS	|||||||	GOSP	
AM	620	WWNR	|||||	NWS, TLK	

CHARLESTON, WV

	FREQ	STATION	POWER	FORMAT	
	94.5	WBES	III	AC	
	99.9	WVAF	|||||	AC	
	102.7	WVSR	|||||	TOP 40	
	105.1	WKLC	|||||||	ROCK	
	107.3	WKAZ	|||||||	OLDS	
	96.1	WKWS	|||||	CTRY	
	97.5	WQBE	|||||	CTRY	
AM	680	WCAW	|||||||	BB, NOS	
	88.5	WVPN	|||||	NWS, TLK, NPR, PRI	P
	100.9	WJYP	III	RLG AC	
AM	580	WCHS	|||||||	NWS, TLK	

38

	FREQ	STATION	POWER	FORMAT	
AM	950	WQBE	IIIIIIII	NWS, TLK	
AM	1490	WSWW	I	NWS, TLK	

WV

HUNTINGTON, WV

	FREQ	STATION	POWER	FORMAT	
	100.5	WKEE	IIIIIIII	HOT AC	
	102.3	WUGO	III	AC	
	106.3	WAMX	III	MOD RK	
	107.1	WFXN	I	CLS RK	
	92.7	WRVC	III	OLDS	
	97.1	WBVB	III	OLDS	
	93.7	WDGG	IIIIIIII	CTRY	
	103.3	WTCR	IIIII	CTRY	
	105.7	WLGC	III	CTRY	
AM	1420	WTCR	IIIII	CTRY	
AM	800	WKEE	IIIIIIII	BB, NOS	
	88.1	WMUL	I	DVRS	C
	89.1	WOUL	IIIII	CLS, NPR, PRI	P
	107.9	WEMM	IIIII	GOSP	
AM	1340	WCMI	I	GOSP	
AM	930	WRVC	IIIIIIII	NWS, TLK	
AM	1230	WIRO	I	SPRTS	

MORGANTOWN, WV

	FREQ	STATION	POWER	FORMAT	
	92.7	WVHF	III	AC	
	104.1	WDCI	III	LGHT AC	
	101.9	WVAQ	IIIII	TOP 40	
	94.3	WRLF	III	OLDS	
	100.1	WCLG	III	CLS RK	
	100.9	WMQC	III	CLS RK	
	106.5	WFBY	IIIII	CLS RK	
	105.7	WOBG	III	OLDS	
AM	1400	WOBG	I	OLDS	
AM	1490	WTCS	I	OLDS	
	93.1	WVUC	III	CTRY	
	97.9	WKKW	IIIII	CTRY	
	102.7	WTUS	III	CTRY	
	104.9	WPDX	III	CTRY	
	91.7	WWVU	I	AC	C
AM	920	WMMN	IIIIIIII	NWS	
AM	1340	WHAR	I	NWS, TLK	
AM	1440	WAJR	IIIII	NWS, TLK	

PARKERSBURG, WV

	FREQ	STATION	POWER	FORMAT	
	102.1	WRVB	III	AC	
	95.1	WXIL	IIIII	TOP 40	
	103.1	WHBR	III	MOD RK	
	100.1	WDMX	III	OLDS	
AM	1050	WADC	IIIII	OLDS	
	99.1	WXKX	III	CTRY	
	107.1	WNUS	III	CTRY	
	88.3	WMRT	I	CLS	C
AM	1230	WKYG	I	KIDS	
AM	1490	WMOA	I	EZ	
AM	1450	WLTP	I	NWS, TLK	

WHEELING, WV

	FREQ	STATION	POWER	FORMAT	
	93.5	WBNV	III	LGHT AC	
	97.3	WKWK	IIIII	AC	
	105.5	WZNW	III	AC	
	100.5	WOMP	III	TOP 40	
	107.5	WEGW	III	ROCK	
	95.7	WEEL	III	OLDS	

	FREQ	STATION	POWER	FORMAT	
	98.7	WOVK	IIIII	CTRY	
AM	1400	WBBD	I	BB, NOS	
	89.9	WVNP	III	NWS, TLK, NPR, PRI	P
	91.5	WGLZ	I	DVRS	C
	96.5	WRKP	III	RLG AC	
AM	1170	WWVA	IIIIIIII	NWS, TLK	

WV

		88.1	
DC	WASH. D.C.	WMUC	18
IL	BLOOMINGTON	WESN	18
IL	CHICAGO	WETN	19
IL	CHICAGO	WLRA	19
IL	CHICAGO	WCRX	19
KY	LEXINGTON	WRFL	22
MD	BALTIMORE	WJHU	23
NJ	TRENTON	WNJT	28
OH	AKRON	WZIP	28
OH	DAYTON	WDPR	30
OH	TOLEDO	WBGU	30
PA	ALLENTOWN	WDIY	31
PA	HARRISBURG	WXPH	32
PA	PHILADELPHIA	WNJS	32
PA	PITTSBURGH	WRSK	33
PA	WILLIAMSPORT	WPTC	34
PA	YORK	WVYC	35
VA	NORFOLK	WHOV	35
WI	GREEN BAY	WHID	37
WV	HUNTINGTON	WMUL	39
		88.3	
IL	CHICAGO	WZRD	19
IN	EVANSVILLE	WNIN	21
MI	ANN ARBOR	WCBN	25
OH	CLEVELAND	WBWC	29
OH	TOLEDO	WXUT	30
PA	HARRISBURG	WDCV	32
PA	LANCASTER	WWEC	32
PA	PITTSBURGH	WRCT	33
WV	PARKERSBURG	WMRT	39
		8.5	
DC	WASH. D.C.	WAMU	18
MI	GRAND RAPIDS	WGVU	26
OH	CINCINNATI	WMUB	28
OH	YOUNGSTOWN	WYSU	30
PA	ERIE	WMCE	31
PA	PHILADELPHIA	WXPN	32
PA	SCRANTON	WRKC	34
VA	CHARLOTSVLL	WVTW	35
WV	CHARLESTON	WVPN	39
		88.7	
IL	CHAMPAIGN	WPCD	18
IL	CHICAGO	WRSE	19
IL	CHICAGO	WLUW	19
IN	INDIANAPOLIS	WICR	21
MI	NW MICHIGAN	WIAA	27
OH	CINCINNATI	WOBO	28
OH	CLEVELAND	WJCU	29
PA	HARRISBURG	WSYC	32
WI	MADISON	WERN	37
		88.9	
IL	CHICAGO	WMXM	19
IL	CHICAGO	WRRG	19
IN	SOUTH BEND	WSND	22
KY	LEXINGTON	WEKU	22
MD	BALTIMORE	WEAA	23
MI	GRAND RAPIDS	WBLU	26
MI	LANSING	WDBM	26
OH	DAYTON	WCSU	30
PA	ERIE	WFSE	31
PA	PHILADELPHIA	WBZC	32
VA	RICHMOND	WCVE	36
WI	LA CROSSE	WLSU	37
		89.1	
IL	BLOOMINGTON	WGLT	18
IL	CHICAGO	WONC	19
IN	FT. WAYNE	WBNI	21
MD	HAGERSTOWN	WETH	24
MI	ANN ARBOR	WEMU	25
MI	KALAMAZOO	WIDR	26
NJ	TRENTON	WWFM	28
OH	CLEVELAND	WKSV	29
PA	LANCASTER	WFNM	32
PA	PHILADELPHIA	WYBF	32
VA	ROANOKE	WVTF	36
WV	HUNTINGTON	WOUL	39
		89.3	
DC	WASH. D.C.	WPFW	18
IL	CHICAGO	WNUR	19
IL	CHICAGO	WKKC	19
KY	LOUISVILLE	WFPL	23
MI	DETROIT	WHFR	25
NJ	MON-OCEN. CTY	WCNJ	27

OH	CLEVELAND	WCSB	29
PA	PITTSBURGH	WQED	33
VA	CHARLOTSVLL	WVTU	35
WI	GREEN BAY	WPNE	37
		89.5	
IN	INDIANAPOLIS	WFCI	21
MD	SALISBURY	WSCL	24
PA	HARRISBURG	WITF	32
VA	NORFOLK	WHRV	35
		89.7	
IN	TERRE HAUTE	WISU	22
MD	BALTIMORE	WTMD	23
MI	LANSING	WLNZ	26
NJ	ATLANTIC CITY	WNJN	27
OH	AKRON	WKSU	28
OH	CINCINNATI	WNKU	28
OH	COLUMBUS	WOSU	29
PA	HARRISBURG	WQEJ	32
PA	PHILADELPHIA	WGLS	32
WI	EAU CLAIRE	WUEC	37
WI	MILWAUKEE	WUWM	38
		89.9	
IL	PEORIA	WCBU	20
KY	LEXINGTON	WRVG	23
MD	FREDERICK	WMTB	24
MI	GRAND RAPIDS	WTHS	26
PA	ERIE	WERG	31
PA	SCRANTON	WVIA	34
VA	ROANOKE	WMRL	36
WI	MADISON	WORT	37
WI	WAUSAU	WWSP	38
WV	WHEELING	WVNP	40
		90.1	
IL	CHAMPAIGN	WEFT	18
IN	INDIANAPOLIS	WFYI	21
MI	SAGINAW	WUCX	27
PA	PHILADELPHIA	WRTI	32
PA	PITTSBURGH	WSRU	33
VA	RICHMOND	WDCE	36
		90.3	
KY	OWENSBORO	WKWC	23
OH	CLEVELAND	WCPN	29
PA	ALLENTOWN	WXLV	31
PA	SCRANTON	WESS	34
VA	NORFOLK	WHRO	35
WI	APPLETON	WRST	37
WI	LA CROSSE	WHLA	37
		90.5	
IL	ROCKFORD	WNIU	20
IN	TERRE HAUTE	WMHD	22
KY	LOUISVILLE	WUOL	23
MI	LANSING	WKAR	26
NJ	MON-OCEN. CTY	WBJB	27
PA	PITTSBURGH	WDUQ	33
		90.7	
MI	NW MICHIGAN	WNMC	27
OH	LIMA	WGLE	30
PA	HARRISBURG	WVMM	32
PA	SCRANTON	WCLH	34
PA	STATE COLL.	WKPS	34
VA	HARRISONBRG	WMRA	35
VA	NORFOLK	WCWM	35
WI	MILWAUKEE	WHAD	38
		90.9	
DC	WASH. D.C.	WETA	18
IL	CHAMPAIGN	WILL	18
IL	CHICAGO	WDCB	19
IN	EVANSVILLE	WKUE	21
OH	CINCINNATI	WGUC	28
PA	PHILADELPHIA	WHYY	32
WI	WAUSAU	WHRM	38
		91.1	
MI	FLINT	WFUM	25
NJ	MORRISTOWN	WWNJ	27
OH	CANTON	WRMU	28
OH	CLEVELAND	WRUW	29
OH	COLUMBUS	WDUB	29
PA	PHILADELPHIA	WRTY	32
PA	SCRANTON	WBUQ	34
PA	YORK	WZBT	35
VA	CHARLOTSVLL	WTJU	35
VA	NORFOLK	WNSB	35
WI	APPLETON	WLFM	37

91.3

DE	WILMINGTON	WVUD	18
IN	BLOOMINGTON	WFHB	20
KY	LEXINGTON	WUKY	23
MD	SALISBURY	WESM	24
MI	DETROIT	WSGR	25
NJ	TRENTON	WTSR	28
OH	DAYTON	WYSO	30
OH	TOLEDO	WGTE	30
PA	ALLENTOWN	WLVR	31
PA	ERIE	WQLN	31
PA	LANCASTER	WLCH	32
PA	PITTSBURGH	WYEP	33
PA	READING	WXAC	33
VA	RICHMOND	WVST	36

91.5

IL	CHICAGO	WBEZ	19
IN	EVANSVILLE	WUEV	21
MD	BALTIMORE	WBJC	24
OH	CLEVELAND	WOBC	29
PA	PHILADELPHIA	WSRN	32
PA	PHILADELPHIA	WDBK	32
PA	SCRANTON	WVMW	34
PA	STATE COLL.	WPSU	34
WV	WHEELING	WGLZ	40

91.7

MI	ANN ARBOR	WUOM	25
NJ	ATLANTIC CITY	WLFR	27
OH	CINCINNATI	WVXU	28
PA	ALLENTOWN	WMUH	31
PA	HARRISBURG	WJAZ	32
PA	LANCASTER	WIXQ	32
PA	PHILADELPHIA	WKDU	32
PA	WILLIAMSPORT	WRLC	34
VA	HARRISONBRG	WEMC	35
WI	MILWAUKEE	WMSE	38
WV	MORGANTOWN	WWVU	39

91.9

DC	WASH. D.C.	WGTS	18
IL	CARBONDALE	WSIU	20
IL	SPRINGFIELD	WUIS	20
KY	LOUISVILLE	WFPK	23
PA	ALLENTOWN	WNTI	31
PA	PITTSBURGH	WVCS	33
VA	CHARLOTSVLL	WNRN	35
WI	WAUSAU	WXPW	38

92.1

MD	HAGERSTOWN	WSRT	24
MD	SALISBURY	WLBW	24
MI	GRAND RAPIDS	WGHN	26
MI	LANSING	WWDX	26
OH	DAYTON	WROU	30
OH	LIMA	WZOQ	30
PA	HARRISBURG	WNCE	32
PA	JOHNSTOWN	WGLU	32
PA	SCRANTON	WQFM	34
VA	NORFOLK	WSVV	36
VA	RICHMOND	WCDX	36
WI	MADISON	WMAD	37
WI	MILWAUKEE	WEZY	38

92.3

IL	CHICAGO	WYCA	19
IL	PEORIA	WBGE	20
IN	BLOOMINGTON	WTTS	20
IN	FT. WAYNE	WFWI	21
MD	BALTIMORE	WERQ	24
MI	DETROIT	WMXD	25
OH	CLEVELAND	WZJM	29
OH	COLUMBUS	WCOL	29
VA	ROANOKE	WXLK	36
WI	WAUSAU	WOSQ	38

92.5

IL	CHAMPAIGN	WKIO	18
KY	OWENSBORO	WBKR	23
OH	CANTON	WZKL	28
OH	CINCINNATI	WOFX	28
OH	TOLEDO	WVKS	30
PA	PHILADELPHIA	WXTU	32
WI	MILWAUKEE	WBWI	38

92.7

DC	WASH. D.C.	WMJS	18
IL	CARBONDALE	WVZA	20
MD	SALISBURY	WGMD	24

MI	FLINT	WDZZ	25
NJ	MON-OCEN. CTY	WOBM	27
PA	YORK	WEGK	35
VA	CHARLOTSVLL	WUVA	35
WV	HUNTINGTON	WRVC	39
WV	MORGANTOWN	WVHF	39

92.9

IN	SOUTH BEND	WNDU	22
KY	LEXINGTON	WVLK	22
OH	DAYTON	WGTZ	30
PA	PITTSBURGH	WLTJ	33
PA	SCRANTON	WMGS	34
VA	NORFOLK	WFOG	35

93.1

IL	CHICAGO	WXRT	19
IN	INDIANAPOLIS	WNAP	21
MD	BALTIMORE	WPOC	23
MI	DETROIT	WDRQ	25
OH	CLEVELAND	WZAK	29
OH	LIMA	WFGF	30
WV	MORGANTOWN	WVUC	39

93.3

IL	PEORIA	WPBG	20
MI	SAGINAW	WKQZ	27
OH	YOUNGSTOWN	WBBG	30
PA	PHILADELPHIA	WMMR	32
PA	WILLIAMSPORT	WHTO	34
WI	LA CROSSE	WIZM	37
WI	MILWAUKEE	WJZI	38

93.5

IN	EVANSVILLE	WJPS	21
IN	LAFAYETTE	WKHY	22
MD	SALISBURY	WZBH	24
MI	DETROIT	WHMI	25
MI	NW MICHIGAN	WBCM	27
OH	TOLEDO	WRQN	30
PA	HARRISBURG	WTPA	31
PA	SCRANTON	WSBG	34
VA	ROANOKE	WJLM	36
WI	APPLETON	WOZZ	36
WV	WHEELING	WBNV	40

93.7

DE	WILMINGTON	WSTW	18
MI	GRAND RAPIDS	WBCT	26
OH	DAYTON	WFCJ	30
PA	PITTSBURGH	WBZZ	33
PA	SCRANTON	WCTP	34
WV	HUNTINGTON	WDGG	39

93.9

DC	WASH. D.C.	WKYS	18
IL	CHICAGO	WLIT	19
IN	INDIANAPOLIS	WGRL	21

94.1

IN	FT. WAYNE	WYSR	21
OH	CANTON	WHBC	28
OH	CINCINNATI	WVMX	28
PA	PHILADELPHIA	WYSP	32
VA	NORFOLK	WXEZ	36
WI	EAU CLAIRE	WIAL	37

94.3

IL	PEORIA	WFXF	20
MD	HAGERSTOWN	WCHA	24
MD	SALISBURY	WICO	24
MI	NW MICHIGAN	WBYB	27
NJ	MON-OCEN. CTY	WJLK	27
PA	SCRANTON	WCTD	34
WI	APPLETON	WROE	36
WV	MORGANTOWN	WRLF	39
IL	CHAMPAIGN	WLRW	18

94.5

KY	LEXINGTON	WMXL	22
MI	GRAND RAPIDS	WKLQ	26
NJ	TRENTON	WNJO	27
OH	DAYTON	WBTT	30
PA	LANCASTER	WDAC	32
PA	PITTSBURGH	WWSW	33
VA	RICHMOND	WRVQ	36
WI	MILWAUKEE	WKTI	38
WV	CHARLESTON	WBES	39

94.7

DC	WASH. D.C.	WARW	18
IL	CHICAGO	WXCD	19
IN	INDIANAPOLIS	WFBQ	21

KY	LOUISVILLE	WLSY	23
KY	OWENSBORO	WBIO	24
MI	DETROIT	WCSX	25
OH	COLUMBUS	WSNY	29
PA	ERIE	WFGO	31
WI	WAUSAU	WOFM	38

94.9

IL	DANVILLE	WRHK	19
MI	LANSING	WMMQ	26
OH	CINCINNATI	WVAE	28
OH	CLEVELAND	WQMX	29
PA	HARRISBURG	WRBT	31
PA	PITTSBURGH	WASP	33
VA	NORFOLK	WPTE	35
VA	ROANOKE	WPVR	36

95.1

IL	CARBONDALE	WXLT	20
IL	CHICAGO	WIIL	19
IN	FT. WAYNE	WAJI	21
MD	BALTIMORE	WRBS	24
MD	HAGERSTOWN	WIKZ	24
MI	FLINT	WFBE	25
NJ	ATLANTIC CITY	WAYV	27
PA	ALLENTOWN	WZZO	31
VA	CHARLOTSVLL	WQMZ	35
WI	EAU CLAIRE	WQRB	37
WV	PARKERSBURG	WXIL	39

95.3

IL	CHAMPAIGN	WZNF	18
IL	ROCKFORD	WKMQ	20
IN	TERRE HAUTE	WNDI	22
MD	SALISBURY	WJNE	24
MI	BATTLE CREEK	WBXX	25
OH	DAYTON	WZLR	30
PA	PITTSBURGH	WJPA	33
PA	STATE COLL.	WZWW	34
VA	RICHMOND	WKHK	36

95.5

DC	WASH. D.C.	WPGC	18
IL	CHICAGO	WNUA	19
IL	PEORIA	WGLO	20
IN	INDIANAPOLIS	WFMS	21
MI	DETROIT	WKQI	25
MI	NW MICHIGAN	WJZJ	26
OH	CLEVELAND	WCLV	29
OH	COLUMBUS	WHOK	29
PA	JOHNSTOWN	WKYE	32
PA	WILLIAMSPORT	WMYL	34
WI	WAUSAU	WIFC	38

95.7

KY	LOUISVILLE	WQMF	23
MI	GRAND RAPIDS	WLHT	26
OH	DAYTON	WCLR	30
PA	PHILADELPHIA	WXXM	32
PA	SCRANTON	WKQV	34
VA	NORFOLK	WVKL	35
WI	LA CROSSE	WRQT	37
WI	MILWAUKEE	WZTR	38
WV	WHEELING	WEEL	40

95.9

IN	INDIANAPOLIS	WPZZ	21
MD	BALTIMORE	WWIN	24
MD	HAGERSTOWN	WYII	24
MD	SALISBURY	WOSC	24
MI	FLINT	WGRI	25
NJ	MON-OCEN. CTY	WRAT	27
WI	GREEN BAY	WKSZ	37

96.1

IL	CHAMPAIGN	WQQB	18
MI	GRAND RAPIDS	WVTI	26
MI	SAGINAW	WHNN	27
NJ	ATLANTIC CITY	WTTH	27
OH	TOLEDO	WMTR	30
PA	ALLENTOWN	WCTO	31
PA	PITTSBURGH	WDRV	33
PA	YORK	WSOX	35
VA	HARRISONBRG	WLTK	35
WV	CHARLESTON	WKWS	39

96.3

DC	WASH. D.C.	WHUR	18
IL	CHICAGO	WBBM	19
IN	FT. WAYNE	WEJE	21
IN	INDIANAPOLIS	WHHH	21

MI	DETROIT	WPLT	25
MI	NW MICHIGAN	WLXT	26
OH	COLUMBUS	WLVQ	29
PA	WILLIAMSPORT	WJSA	34

96.5

IN	LAFAYETTE	WAZY	22
KY	LOUISVILLE	WGZB	23
MI	KALAMAZOO	WFAT	26
MI	LANSING	WQHH	26
OH	AKRON	WKDD	28
OH	CINCINNATI	WYGY	28
PA	JOHNSTOWN	WMTZ	32
PA	PHILADELPHIA	WWDB	33
WI	MILWAUKEE	WKLH	38
WV	WHEELING	WRKP	40

96.7

IL	BLOOMINGTON	WIHN	18
IL	ROCKFORD	WLUV	20
IN	BLOOMINGTON	WBWB	20
MD	HAGERSTOWN	WQCM	24
MI	BATTLE CREEK	WUFN	25
WI	WAUSAU	WYTE	38

96.9

IL	CHICAGO	WNIZ	19
KY	LEXINGTON	WGKS	22
MD	SALISBURY	WBEY	24
MI	GRAND RAPIDS	WLAV	26
NJ	ATLANTIC CITY	WFPG	27
OH	DAYTON	WRNB	30
PA	LANCASTER	WLAN	32
PA	PITTSBURGH	WRRK	33
WI	APPLETON	WUSW	37

97.1

DC	WASH. D.C.	WASH	18
IL	CHICAGO	WNIB	19
IN	INDIANAPOLIS	WENS	21
MD	SALISBURY	WQJZ	24
MI	DETROIT	WKRK	25
OH	COLUMBUS	WBNS	29
PA	SCRANTON	WBHT	34
PA	STATE COLL.	WQWK	34
WV	HUNTINGTON	WBVB	39

97.3

IL	PEORIA	WFYR	20
IN	FT. WAYNE	WMEE	21
MI	SAGINAW	WIXC	27
OH	CINCINNATI	WYLX	28
PA	HARRISBURG	WRVV	31
VA	NORFOLK	WGH	35
WI	MILWAUKEE	WLTQ	38
WV	WHEELING	WKWK	40

97.5

IL	CHAMPAIGN	WHMS	18
IL	ROCKFORD	WZOK	20
KY	LOUISVILLE	WAMZ	23
MI	LANSING	WJIM	26
MI	NW MICHIGAN	WKLT	26
NJ	TRENTON	WPST	27
OH	AKRON	WONE	28
VA	CHARLOTSVLL	WWWV	35
WV	CHARLESTON	WQBE	39

97.7

IL	CARBONDALE	WQUL	20
IN	TERRE HAUTE	WSDM	22
MD	SALISBURY	WAFL	24
OH	CINCINNATI	WOXY	28
PA	JOHNSTOWN	WSGY	32
PA	WILLIAMSPORT	WVRT	34

97.9

IL	CHICAGO	WLUP	19
MD	BALTIMORE	WIYY	23
MD	SALISBURY	WSBL	24
MI	DETROIT	WJLB	25
MI	GRAND RAPIDS	WGRD	26
OH	COLUMBUS	WNCI	29
PA	ERIE	WXTA	31
PA	SCRANTON	WZMT	34
VA	ROANOKE	WRVX	36
WI	WAUSAU	WSPT	38
WV	MORGANTOWN	WKKW	39

98.1

KY	LEXINGTON	WBUI	22
MI	SAGINAW	WKCQ	27

OH	CANTON	WHK	28
PA	ALTOONA	WFGY	31
PA	PHILADELPHIA	WOGL	32
VA	RICHMOND	WTVR	36
WI	EAU CLAIRE	WISM	37
WI	MADISON	WMGN	37

98.3

IL	CHICAGO	WCCQ	19
IN	INDIANAPOLIS	WXIR	22
PA	PITTSBURGH	WZKT	33
WI	MILWAUKEE	WFMR	38

98.5

IL	PEORIA	WEEK	20
IN	TERRE HAUTE	WACF	22
MD	SALISBURY	WGBG	24
NJ	MON-OCEN. CTY	WBBO	27
OH	CINCINNATI	WRRM	28
OH	CLEVELAND	WNCX	29
PA	SCRANTON	WKRZ	34
PA	YORK	WYCR	35
VA	HARRISONBRG	WACL	35
WI	GREEN BAY	WQLH	37

98.7

DC	WASH. D.C.	WMZQ	18
IL	CHICAGO	WFMT	19
IL	SPRINGFIELD	WNNS	20
IN	LAFAYETTE	WASK	22
MI	DETROIT	WVMV	25
MI	GRAND RAPIDS	WFGR	26
OH	COLUMBUS	WSLN	29
VA	NORFOLK	WNOR	35
WV	WHEELING	WOVK	40

98.9

MI	NW MICHIGAN	WKLZ	26
OH	COLUMBUS	WMXG	29
OH	YOUNGSTOWN	WKBN	30
PA	HARRISBURG	WQLV	31
PA	PHILADELPHIA	WUSL	33

99.1

IL	DANVILLE	WIAI	19
MI	LANSING	WFMK	26
OH	DAYTON	WHKO	30
PA	JOHNSTOWN	WQKK	32
VA	NORFOLK	WXGM	35
VA	ROANOKE	WSLQ	36
WI	MILWAUKEE	WMYX	38
WV	PARKERSBURG	WXKX	39

99.3

MI	GRAND RAPIDS	WJQK	26
MI	NW MICHIGAN	WBNZ	26
NJ	ATLANTIC CITY	WSAX	27
OH	CINCINNATI	WSCH	28
PA	HARRISBURG	WWKL	31
PA	PITTSBURGH	WPQR	33
PA	WILLIAMSPORT	WZXR	34
VA	RICHMOND	WPLZ	36

99.5

DC	WASH. D.C.	WGAY	18
DE	WILMINGTON	WJBR	18
IL	CHICAGO	WUSN	19
IN	EVANSVILLE	WKDQ	21
IN	INDIANAPOLIS	WZPL	21
MI	DETROIT	WYCD	25
OH	CLEVELAND	WGAR	29
PA	SCRANTON	WUSR	34
WV	BECKLEY	WJLS	38

99.7

KY	LOUISVILLE	WDJX	23
OH	COLUMBUS	WBZX	29
PA	PITTSBURGH	WSHH	33

99.9

IL	CARBONDALE	WOOZ	20
IL	PEORIA	WIXO	20
IN	TERRE HAUTE	WTHI	22
MD	FREDERICK	WFRE	24
MD	SALISBURY	WWFG	24
OH	DAYTON	WLQT	30
OH	TOLEDO	WKKO	30
PA	ALLENTOWN	WODE	31
PA	ERIE	WXKC	31
WI	WAUSAU	WIZD	38
WV	CHARLESTON	WVAF	39

100.1

KY	LEXINGTON	WKQQ	22
NJ	MON-OCEN. CTY	WJRZ	27
OH	AKRON	WNIR	28
PA	ALTOONA	WPRR	31
PA	HARRISBURG	WQIC	31
VA	ROANOKE	WLYK	36
WI	LA CROSSE	WKBH	37
WV	MORGANTOWN	WCLG	39
WV	PARKERSBURG	WDMX	39

100.3

DC	WASH. D.C.	WBIG	18
IL	CHAMPAIGN	WIXY	18
IL	CHICAGO	WNND	19
MI	DETROIT	WNIC	25
OH	COLUMBUS	WCLT	29
PA	PHILADELPHIA	WPLY	32
VA	RICHMOND	WSOJ	36
WI	APPLETON	WNCY	37

100.5

KY	LOUISVILLE	WTFX	23
MI	GRAND RAPIDS	WTRV	26
MI	SAGINAW	WTCF	27
PA	HARRISBURG	WYGL	31
VA	NORFOLK	WCMS	35
WV	HUNTINGTON	WKEE	39
WV	WHEELING	WOMP	40

100.7

IN	TERRE HAUTE	WMGI	22
MD	BALTIMORE	WGRX	23
MI	LANSING	WITL	26
OH	CLEVELAND	WMMS	29
PA	ALLENTOWN	WLEV	31
PA	PITTSBURGH	WZPT	33
VA	HARRISONBRG	WQPO	35
WI	EAU CLAIRE	WBIZ	37
WI	MILWAUKEE	WKKV	38

100.9

IL	DANVILLE	WHPO	19
IL	ROCKFORD	WQFL	20
IN	INDIANAPOLIS	WYJZ	21
MI	NW MICHIGAN	WIZY	27
MI	SAGINAW	WMJK	27
OH	CINCINNATI	WIZF	28
PA	ERIE	WRKT	31
WV	CHARLESTON	WJYP	39
WV	MORGANTOWN	WMQC	39

101.1

DC	WASH. D.C.	WWDC	18
IL	CHICAGO	WKQX	19
IN	FT. WAYNE	WLZQ	21
MI	DETROIT	WRIF	25
OH	COLUMBUS	WWCD	29
OH	YOUNGSTOWN	WHOT	30
PA	ALTOONA	WGMR	31
PA	PHILADELPHIA	WBEB	32
VA	RICHMOND	WJRV	36
WI	GREEN BAY	WIXX	37

101.3

IN	LAFAYETTE	WBAA	22
KY	LOUISVILLE	WMJM	23
MD	SALISBURY	WXPZ	24
MI	GRAND RAPIDS	WCUZ	26
PA	LANCASTER	WROZ	32
PA	SCRANTON	WGGY	34
VA	NORFOLK	WWDE	35

101.5

IL	BLOOMINGTON	WBNQ	18
IL	CARBONDALE	WCIL	20
IN	SOUTH BEND	WNSN	22
KY	LEXINGTON	WLRO	22
MD	HAGERSTOWN	WAYZ	24
NJ	TRENTON	WKXW	27
OH	TOLEDO	WRVF	30
PA	PITTSBURGH	WORD	33
VA	ROANOKE	WZZI	36
WI	MADISON	WIBA	37

101.7

DE	WILMINGTON	WJKS	18
IN	FT. WAYNE	WLDE	21
KY	LOUISVILLE	WTHQ	23
MD	SALISBURY	WRBG	24
MI	LANSING	WHZZ	26
OH	COLUMBUS	WNKO	29

| PA | JOHNSTOWN | WSRA | 32 |
| VA | ROANOKE | WJJX | 36 |

101.9

IL	CHICAGO	WTMX	18
IL	SPRINGFIELD	WQQL	20
IN	INDIANAPOLIS	WQFE	21
MD	BALTIMORE	WLIF	23
MI	DETROIT	WDET	25
MI	NW MICHIGAN	WLDR	26
OH	CINCINNATI	WKRQ	28
OH	YOUNGSTOWN	WBTJ	30
VA	CHARLOTSVLL	WVSY	35
WI	WAUSAU	WDEZ	38
WV	MORGANTOWN	WVAQ	39

102.1

IL	DANVILLE	WDNL	19
MI	KALAMAZOO	WMUK	26
OH	CLEVELAND	WDOK	29
OH	LIMA	WIMT	30
PA	PHILADELPHIA	WIOQ	32
VA	RICHMOND	WRXL	36
WI	MILWAUKEE	WLUM	38
WV	PARKERSBURG	WRVB	39

102.3

DC	WASH. D.C.	WMMJ	18
IL	CHICAGO	WXLC	19
IL	PEORIA	WTAZ	20
IN	FT. WAYNE	WGL	21
IN	INDIANAPOLIS	WCBK	21
IN	SOUTH BEND	WGTC	22
KY	LOUISVILLE	WLRS	23
PA	ERIE	WJET	31
PA	HARRISBURG	WHYL	31
PA	SCRANTON	WILK	34
VA	CHARLOTSVLL	WVAO	35
WV	HUNTINGTON	WUGO	39

102.5

KY	LEXINGTON	WLTO	22
MD	SALISBURY	WOLC	24
MI	SAGINAW	WIOG	27
PA	PITTSBURGH	WDVE	33
PA	READING	WRFY	33
WI	MADISON	WNWC	37

IL	CHICAGO	WVAZ	19
IN	TERRE HAUTE	WLEZ	22
MD	BALTIMORE	WXYV	23

102.7

OH	CINCINNATI	WEBN	28
PA	WILLIAMSPORT	WKSB	34
VA	ROANOKE	WLDJ	36
WV	CHARLESTON	WVSR	39
WV	MORGANTOWN	WTUS	39

102.9

IN	FT. WAYNE	WEXI	21
MD	ANN ARBOR	WIQB	25
MI	GRAND RAPIDS	WFUR	26
OH	DAYTON	WING	30
PA	PHILADELPHIA	WMGK	32
VA	NORFOLK	WOWI	35
WI	MILWAUKEE	WLZR	38

103.1

IL	ROCKFORD	WRWC	20
IN	EVANSVILLE	WGBF	21
IN	SOUTH BEND	WHME	22
KY	LOUISVILLE	WRKA	23
MD	BALTIMORE	WRNR	23
MD	FREDERICK	WAFY	24
OH	COLUMBUS	WSMZ	29
PA	SCRANTON	WILP	34
PA	STATE COLL.	WBHV	34
WI	APPLETON	WOGB	37
WV	PARKERSBURG	WHBR	39

103.3

IN	INDIANAPOLIS	WRZX	21
KY	LEXINGTON	WXZZ	22
MI	KALAMAZOO	WKFR	26
PA	YORK	WARM	35
WI	WAUSAU	WGLX	38
WV	HUNTINGTON	WTCR	39

103.5

| DC | WASH. D.C. | WGMS | 18 |
| IL | CARBONDALE | WUEZ | 20 |

IL	CHICAGO	WRCX	19
MD	SALISBURY	WJYN	24
MI	DETROIT	WMUZ	25
MI	NW MICHIGAN	WTCM	27
OH	CINCINNATI	WGRR	28
OH	COLUMBUS	WJZA	29
PA	SCRANTON	WKAB	34
VA	CHARLOTSVLL	WMRY	35

103.7

IL	SPRINGFIELD	WDBR	20
IN	BLOOMINGTON	WFIU	20
MD	BALTIMORE	WXCY	23
NJ	ATLANTIC CITY	WMGM	27
PA	ERIE	WRTS	31
VA	RICHMOND	WMXB	36
WI	MILWAUKEE	WXSS	38
WV	BECKLEY	WCIR	38

103.9

IL	CARBONDALE	WXAN	20
IL	CHICAGO	WZCH	19
IN	FT. WAYNE	WXKE	21
IN	SOUTH BEND	WRBR	22
KY	LOUISVILLE	WMHX	23
MD	FREDERICK	WWVZ	24
MD	SALISBURY	WOCQ	24
MI	NW MICHIGAN	WCMW	27
OH	TOLEDO	WXEG	30
PA	ALTOONA	WALY	31
PA	PHILADELPHIA	WPHI	33
PA	PITTSBURGH	WLSW	33
WI	APPLETON	WVBO	37

104.1

DC	WASH. D.C.	WWZZ	18
IL	BLOOMINGTON	WBWN	18
IN	EVANSVILLE	WIKY	21
MI	GRAND RAPIDS	WVGR	26
OH	CLEVELAND	WQAL	29
PA	ALLENTOWN	WAEB	31
PA	HARRISBURG	WNNK	31
WI	MADISON	WZEE	37
WV	MORGANTOWN	WDCI	39

104.3

IL	CHICAGO	WJMK	19
IN	TERRE HAUTE	WCBH	22
MD	BALTIMORE	WOCT	23
MI	DETROIT	WOMC	25
OH	CINCINNATI	WNLT	28
OH	COLUMBUS	WZJZ	29
VA	HARRISONBRG	WKCY	35

104.5

IL	SPRINGFIELD	WFMB	20
IN	INDIANAPOLIS	WGLD	21
KY	LEXINGTON	WLKT	22
MI	SAGINAW	WMJA	27
PA	PHILADELPHIA	WYXR	32
VA	NORFOLK	WNVZ	35
WI	EAU CLAIRE	WAXX	37

104.7

MD	HAGERSTOWN	WWMD	24
MD	SALISBURY	WQHQ	24
OH	DAYTON	WTUE	30
OH	TOLEDO	WIOT	30
PA	PITTSBURGH	WJJJ	33
WI	MILWAUKEE	WEXT	38

104.9

IL	PEORIA	WXCL	20
IL	ROCKFORD	WXRX	20
MI	BATTLE CREEK	WWKN	25
NJ	ATLANTIC CITY	WRDR	27
OH	CLEVELAND	WZLE	29
OH	LIMA	WAJC	30
PA	ALTOONA	WMXV	31
PA	SCRANTON	WWDL	34
VA	ROANOKE	WRDJ	36
WI	APPLETON	WPCK	37
WI	LA CROSSE	WLXR	37
WI	MILWAUKEE	WTKM	38
WI	WAUSAU	WKQH	38
WV	MORGANTOWN	WPDX	39

105.1

DC	WASH. D.C.	WAVA	18
IL	CARBONDALE	WTAO	20
IL	CHICAGO	WOJO	19

IN	BLOOMINGTON	WGCT	21
IN	FT. WAYNE	WQHK	21
KY	LOUISVILLE	WXLN	23
MI	DETROIT	WXDG	25
OH	CINCINNATI	WUBE	28
PA	LANCASTER	WIOV	32
PA	WILLIAMSPORT	WILQ	34
VA	HARRISONBRG	WAMM	35
WI	MADISON	WYZM	37
WV	CHARLESTON	WKLC	39

105.3

IN	EVANSVILLE	WYNG	21
IN	LAFAYETTE	WKOA	22
KY	LOUISVILLE	WMPI	23
PA	PHILADELPHIA	WDAS	33
VA	NORFOLK	WJCD	35

105.5

IL	CHICAGO	WZSR	19
IL	CHICAGO	WLJE	19
IN	TERRE HAUTE	WWVR	22
MD	SALISBURY	WLVW	24
MI	FLINT	WWCK	25
NJ	MORRISTOWN	WDHA	27
OH	TOLEDO	WWWM	30
VA	ROANOKE	WKDE	36
WI	MADISON	WMMM	37
WV	WHEELING	WZNW	40

105.7

IL	PEORIA	WWCT	20
IN	INDIANAPOLIS	WTLC	21
MD	BALTIMORE	WQSR	23
MI	GRAND RAPIDS	WOOD	26
OH	CLEVELAND	WMJI	29
OH	COLUMBUS	WZAZ	29
OH	COLUMBUS	WXMG	29
PA	JOHNSTOWN	WFJY	32
PA	YORK	WQXA	35
VA	RICHMOND	WDYL	36
WI	APPLETON	WAPL	37
WI	EAU CLAIRE	WCFW	37
WV	HUNTINGTON	WLGC	39
WV	MORGANTOWN	WOBG	39

105.9

DC	WASH. D.C.	WJZW	18
IL	CHAMPAIGN	WGKC	18
IL	CHICAGO	WCKG	19
IN	TERRE HAUTE	WMMC	22
KY	LEXINGTON	WVRB	23
KY	LOUISVILLE	WRVI	23
MD	SALISBURY	WXJN	24
MI	DETROIT	WCHB	25
MI	NW MICHIGAN	WKHQ	26
PA	PITTSBURGH	WXDX	33
PA	STATE COLL.	WUBZ	34
VA	ROANOKE	WLNI	36

106.1

IN	EVANSVILLE	WDKS	21
OH	YOUNGSTOWN	WNCD	30
PA	PHILADELPHIA	WJJZ	32
VA	ROANOKE	WJJS	36
WI	MILWAUKEE	WMIL	38

106.3

IL	CARBONDALE	WQRL	20
IL	CHICAGO	WYBA	19
IN	FT. WAYNE	WSHI	21
IN	SOUTH BEND	WUBU	22
KY	LEXINGTON	WJMM	23
MI	SAGINAW	WGER	27
NJ	MON-OCEN. CTY	WHTG	27
PA	ERIE	WCTL	31
WI	LA CROSSE	WQCC	37
WI	MADISON	WWQM	37
WV	HUNTINGTON	WAMX	39

106.5

MD	BALTIMORE	WWMX	23
MD	SALISBURY	WKHW	24
MI	KALAMAZOO	WQLR	26
OH	CINCINNATI	WNKR	28
OH	CLEVELAND	WMVX	29
OH	TOLEDO	WBUZ	30
PA	SCRANTON	WHLM	34
VA	RICHMOND	WBZU	36
WI	WAUSAU	WLJY	38

WV	MORGANTOWN	WFBY	39

106.7

DC	WASH. D.C.	WJFK	18
IL	CHICAGO	WYLL	19
IN	INDIANAPOLIS	WBKS	22
IN	LAFAYETTE	WGLM	22
KY	LEXINGTON	WKXO	22
MI	DETROIT	WWWW	25
PA	HARRISBURG	WRKZ	31
PA	PITTSBURGH	WAMO	33
WI	GREEN BAY	WJLW	37

106.9

IL	PEORIA	WSWT	20
KY	LOUISVILLE	WVEZ	23
MD	HAGERSTOWN	WARX	24
MD	SALISBURY	WRXS	24
OH	CANTON	WRQK	28
PA	SCRANTON	WEZX	34
VA	NORFOLK	WAFX	35
VA	ROANOKE	WLQE	36
WI	MILWAUKEE	WPNT	38

107.1

IL	CHAMPAIGN	WPGU	18
IN	EVANSVILLE	WBNL	21
IN	INDIANAPOLIS	WSYW	21
MD	ANN ARBOR	WQKL	25
MI	DETROIT	WSAQ	25
MI	SAGINAW	WTLZ	27
NJ	MON-OCEN. CTY	WWZY	27
OH	CINCINNATI	WKFS	28
OH	COLUMBUS	WAZU	29
OH	LIMA	WDOH	30
PA	ALLENTOWN	WWYY	31
PA	PITTSBURGH	WSSZ	33
WV	HUNTINGTON	WFXN	39
WV	PARKERSBURG	WNUS	39

107.3

DC	WASH. D.C.	WRQX	18
IL	CARBONDALE	WDDD	20
OH	CLEVELAND	WNWV	29
OH	TOLEDO	WJUC	30
WV	CHARLESTON	WKAZ	39

107.5

IL	CHICAGO	WGCI	19
IN	EVANSVILLE	WABX	21
IN	TERRE HAUTE	WZZQ	22
MI	DETROIT	WGPR	25
MI	NW MICHIGAN	WCCW	26
OH	CINCINNATI	WIOK	28
OH	COLUMBUS	WCKX	29
OH	LIMA	WBUK	30
PA	READING	WBYN	33
VA	CHARLOTSVLL	WUMX	35
WV	WHEELING	WEGW	40

107.7

KY	LOUISVILLE	WSFR	23
MI	KALAMAZOO	WRKR	26
OH	DAYTON	WMMX	30
OH	TOLEDO	WHMQ	30
PA	SCRANTON	WEMR	34
PA	YORK	WGTY	35

107.9

IL	CHICAGO	WLEY	19
IN	FT. WAYNE	WJFX	21
IN	INDIANAPOLIS	WTPI	21
MI	FLINT	WCRZ	25
OH	CLEVELAND	WENZ	29
OH	COLUMBUS	WXST	29
PA	PITTSBURGH	WDSY	33
PA	SCRANTON	WKRF	34
PA	STATE COLL.	WIKN	34
PA	WILLIAMSPORT	WSFT	34
VA	ROANOKE	WYYD	36
WI	WAUSAU	WYCO	38
WV	HUNTINGTON	WEMM	39

AM

550

OH	CINCINNATI	WKRC	28
PA	SCRANTON	WJMW	34
VA	HARRISONBRG	WSVA	35
WI	WAUSAU	WSAU	38

560

IL	CHICAGO	WIND	19
PA	PHILADELPHIA	WFIL	33
WV	BECKLEY	WJLS	38

570

DC	WASH. D.C.	WWRC	18
OH	YOUNGSTOWN	WKBN	30

580

IL	CHAMPAIGN	WILL	18
MI	NW MICHIGAN	WTCM	27
PA	HARRISBURG	WHP	32
WI	LA CROSSE	WKTY	37
WV	CHARLESTON	WCHS	39

590

KY	LEXINGTON	WVLK	22
MI	KALAMAZOO	WKZO	26
PA	PITTSBURGH	WMBS	33
PA	SCRANTON	WARM	34
VA	ROANOKE	WLVA	36

600

MD	BALTIMORE	WCAO	24
MI	FLINT	WSNL	25

610

OH	COLUMBUS	WTVN	29
PA	PHILADELPHIA	WIP	33
VA	ROANOKE	WSLC	36

620

KY	LOUISVILLE	WTMT	23
WI	MILWAUKEE	WTMJ	38
WV	WINNER	WWNR	38

630

DC	WASH. D.C.	WMAL	18
KY	LEXINGTON	WLAP	23

640

IN	TERRE HAUTE	WBOW	22
MI	GRAND RAPIDS	WMFN	26
OH	AKRON	WHLO	28

670

IL	CHICAGO	WMAQ	19

680

KY	LOUISVILLE	WNAI	23
MD	BALTIMORE	WCBM	24
WI	EAU CLAIRE	WOGO	37
WV	CHARLESTON	WCAW	39

690

MI	DETROIT	WNZK	25

700

OH	CINCINNATI	WLW	28

720

IL	CHICAGO	WGN	19

760

MI	DETROIT	WJR	25

780

IL	CHICAGO	WBBM	19

790

KY	LOUISVILLE	WWKY	23
MI	SAGINAW	WSGW	27
PA	ALLENTOWN	WAEB	31
VA	NORFOLK	WNIS	36
WI	EAU CLAIRE	WAYY	37

800

MD	HAGERSTOWN	WCHA	24
WV	HUNTINGTON	WKEE	39

810

IL	CARBONDALE	WDDD	20

820

MD	FREDERICK	WXTR	24
OH	COLUMBUS	WOSU	29
VA	RICHMOND	WGGM	36

840

KY	LOUISVILLE	WHAS	23

850

IL	CHICAGO	WAIT	19
OH	CLEVELAND	WRMR	29
PA	JOHNSTOWN	WJAC	32
PA	READING	WEEU	33
VA	NORFOLK	WTAR	36

860

IN	EVANSVILLE	WSON	21
MD	BALTIMORE	WBGR	24
PA	PHILADELPHIA	WTEL	32
PA	PITTSBURGH	WAMO	33

870

880

MI	LANSING	WKAR	26

880

OH	COLUMBUS	WRFD	29
VA	ROANOKE	WVLR	36

890

IL	CHICAGO	WLS	19

890

KY	LOUISVILLE	WFIA	23
MD	SALISBURY	WJWL	24

900

MI	FLINT	WFDF	25
PA	SCRANTON	WGBI	34
PA	YORK	WSBA	35
VA	RICHMOND	WRNL	36

920

IN	LAFAYETTE	WBAA	22
NJ	TRENTON	WCHR	28
OH	COLUMBUS	WMNI	29
WI	MILWAUKEE	WOKY	38
WV	MORGANTOWN	WMMN	39

930

MD	FREDERICK	WFMD	24
MI	BATTLE CREEK	WBCK	25
OH	CLEVELAND	WEOL	29
WI	WAUSAU	WLBL	38
WV	HUNTINGTON	WRVC	39

950

IL	CHICAGO	WIDB	19
IN	INDIANAPOLIS	WXLW	22
MI	DETROIT	WWJ	25
PA	PHILADELPHIA	WPEN	32
VA	RICHMOND	WXGI	36
WV	CHARLESTON	WQBE	39

960

IN	SOUTH BEND	WSBT	22
MD	SALISBURY	WTGM	24
PA	HARRISBURG	WHYL	32
VA	ROANOKE	WFIR	36

970

IL	SPRINGFIELD	WMAY	20
KY	LOUISVILLE	WLKY	23
PA	PITTSBURGH	WWSW	33
WI	MADISON	WHA	37

980

DC	WASH. D.C.	WTEM	18
IL	DANVILLE	WITY	20
OH	DAYTON	WONE	30
PA	SCRANTON	WILK	34

990

OH	CANTON	WTIG	28
PA	PHILADELPHIA	WZZD	32

1000

IL	CHICAGO	WMVP	19

1010

VA	NORFOLK	WPMH	36

1020

PA	PITTSBURGH	KDKA	33

1050

MI	ANN ARBOR	WTKA	25
VA	NORFOLK	WCMS	35
WV	PARKERSBURG	WADC	39

1060

PA	PHILADELPHIA	KYW	33

1070

IN	INDIANAPOLIS	WIBC	22
VA	CHARLOTSVLL	WINA	35
WI	MADISON	WTSO	37
WV	BECKLEY	WIWS	38

1080

KY	LOUISVILLE	WKJK	23

1090

MD	BALTIMORE	WBAL	24

1100

OH	CLEVELAND	WTAM	29

1130

MI	DETROIT	WDFN	25
WI	MILWAUKEE	WISN	38

1140

MI	GRAND RAPIDS	WKWM	26
VA	RICHMOND	WRVA	36

1150

DE	WILMINGTON	WDEL	18
OH	LIMA	WIMA	30

47

WI	APPLETON	WHBY	37	VA	NORFOLK	WGH	36
				WI	MADISON	WIBA	38

11650

IL	CHICAGO	WSCR	19
MI	FLINT	WWON	25
NJ	MON-OCEN. CTY	WOBM	27
OH	CINCINNATI	WBOB	28
PA	ALLENTOWN	WYNS	31

1320

MI	LANSING	WILS	26
OH	CINCINNATI	WCVG	28
OH	CLEVELAND	WOBL	29
PA	ALLENTOWN	WTKZ	31
PA	PITTSBURGH	WJAS	33
PA	YORK	WGET	35
WI	WAUSAU	WFHR	38

1170

WV	WHEELING	WWVA	40

1190

IN	FT. WAYNE	WOWO	21
WI	MADISON	WNWC	37

1330

IN	EVANSVILLE	WVHI	21
MD	BALTIMORE	WASA	23
MI	FLINT	WTRX	25
OH	CLEVELAND	WELW	29
PA	ERIE	WFLP	31

1200

IL	CHICAGO	WLXX	19

1210

PA	PHILADELPHIA	WPHT	33

1340

DC	WASH. D.C.	WYCB	18
IL	CARBONDALE	WJPF	20
KY	LEXINGTON	WEKY	22
MI	GRAND RAPIDS	WBBL	26
MI	NW MICHIGAN	WMBN	27
NJ	ATLANTIC CITY	WMID	27
OH	DAYTON	WIZE	30
PA	ALTOONA	WTRN	31
PA	PHILADELPHIA	WHAT	33
PA	PITTSBURGH	WCVI	33
PA	READING	WRAW	33
PA	SCRANTON	WYCK	34
PA	WILLIAMSPORT	WWPA	34
WI	MILWAUKEE	WJYI	38
WV	HUNTINGTON	WCMI	39
WV	MORGANTOWN	WHAR	39

1220

OH	CLEVELAND	WKNR	29

1230

IL	BLOOMINGTON	WJBC	18
IL	CHICAGO	WJOB	19
MD	BALTIMORE	WITH	24
MI	GRAND RAPIDS	WTKG	26
OH	CINCINNATI	WUBE	28
OH	COLUMBUS	WFII	29
OH	TOLEDO	WCWA	30
PA	ALLENTOWN	WEEX	31
PA	HARRISBURG	WKBO	32
PA	PITTSBURGH	WBVP	33
VA	NORFOLK	WNOR	35
WI	WAUSAU	WXCO	38
WV	HUNTINGTON	WIRO	39
WV	PARKERSBURG	WKYG	40

1350

IL	PEORIA	WOAM	20
KY	LOUISVILLE	WLOU	23
NJ	TRENTON	WHWH	28
OH	AKRON	WTOU	28
PA	YORK	WOYK	35
VA	NORFOLK	WGPL	35
VA	ROANOKE	WBLT	36

1240

IL	SPRINGFIELD	WTAX	20
KY	LOUISVILLE	WLLV	23
MD	HAGERSTOWN	WJEJ	24
MI	LANSING	WJIM	26
OH	YOUNGSTOWN	WBBW	30
PA	ALTOONA	WRTA	31
PA	READING	WIOV	33
PA	SCRANTON	WBAX	34
VA	RICHMOND	WGCV	36
VA	ROANOKE	WGMN	36

1360

MD	BALTIMORE	WWLG	23
MI	KALAMAZOO	WKMI	26
OH	CINCINNATI	WCKY	28
PA	PITTSBURGH	WPTT	33
VA	HARRISONBRG	WHBG	35
WI	GREEN BAY	WGEE	37

1250

IN	FT. WAYNE	WGL	21
NJ	MORRISTOWN	WMTR	27
PA	PITTSBURGH	WEAE	33
PA	YORK	WQXA	35
WI	MILWAUKEE	WEMP	38

1370

IN	BLOOMINGTON	WGCL	21
OH	TOLEDO	WSPD	30

1380

IL	ROCKFORD	WTJK	20
IN	FT. WAYNE	WHWD	21
MI	DETROIT	WPHM	25
VA	RICHMOND	WTVR	36

1260

DC	WASH. D.C.	WWDC	18
IN	INDIANAPOLIS	WNDE	22
MI	GRAND RAPIDS	WWJQ	26
NJ	TRENTON	WBUD	28
OH	CLEVELAND	WMIH	29
PA	ERIE	WRIE	31
VA	CHARLOTSVLL	WCHV	35

1390

DC	WASH. D.C.	WZHF	18
IL	CHICAGO	WGCI	19
OH	YOUNGSTOWN	WRTK	30
PA	LANCASTER	WLAN	32
PA	STATE COLL.	WRSC	34
WI	WAUSAU	WRIG	38

1270

MI	DETROIT	WXYT	25
PA	HARRISBURG	WLBR	31
VA	NORFOLK	WTJZ	35

1400

IL	CHAMPAIGN	WDWS	18
IN	EVANSVILLE	WEOA	21
MD	BALTIMORE	WWIN	24
MI	BATTLE CREEK	WRCC	25
MI	DETROIT	WQBH	25
MI	SAGINAW	WSAM	27
NJ	ATLANTIC CITY	WOND	27
PA	ALLENTOWN	WEST	31
PA	ERIE	WLKK	31
PA	HARRISBURG	WTCY	32
PA	SCRANTON	WICK	34
PA	WILLIAMSPORT	WRAK	34
VA	CHARLOTSVLL	WKAV	35
VA	NORFOLK	WPCE	36
WI	EAU CLAIRE	WBIZ	37
WI	GREEN BAY	WDUZ	37
WI	MILWAUKEE	WRJN	38
WV	MORGANTOWN	WOBG	39
WV	WHEELING	WBBD	40

1280

IN	EVANSVILLE	WGBF	21
PA	YORK	WHVR	35
WI	APPLETON	WNAM	37

1290

IL	PEORIA	WIRL	20
OH	DAYTON	WHIO	30
PA	ALTOONA	WFBG	31
WI	MILWAUKEE	WMCS	38

1300

IL	CHICAGO	WTAQ	19
KY	LEXINGTON	WLXG	23
MD	BALTIMORE	WJFK	24
MI	GRAND RAPIDS	WOOD	26
NJ	TRENTON	WIMG	28
OH	CLEVELAND	WERE	29

1310

IN	INDIANAPOLIS	WTLC	21
MI	DETROIT	WYUR	25
NJ	MON-OCEN. CTY	WADB	27
OH	CANTON	WDPN	28

	1410		
OH	DAYTON	WING	30
PA	ALLENTOWN	WLSH	31
PA	PITTSBURGH	KQV	33
WI	LA CROSSE	WIZM	37

	1420		
IL	CARBONDALE	WINI	20
KY	OWENSBORO	WVJS	23
MI	FLINT	WFLT	25
MI	KALAMAZOO	WKPR	26
OH	CLEVELAND	WHK	29
WV	HUNTINGTON	WTCR	39

	1430		
IN	INDIANAPOLIS	WMYS	21
PA	ALTOONA	WVAM	31

	1440		
IL	ROCKFORD	WROK	20
IN	TERRE HAUTE	WPRS	22
MI	SAGINAW	WMAX	27
OH	YOUNGSTOWN	WRRO	30
WI	GREEN BAY	WNFL	37
WV	MORGANTOWN	WAJR	39

	1450		
DC	WASH. D.C.	WOL	18
DE	WILMINGTON	WILM	18
IL	CHICAGO	WVON	19
IL	SPRINGFIELD	WFMB	20
IN	LAFAYETTE	WASK	22
KY	LOUISVILLE	WAVG	23
MI	GRAND RAPIDS	WHTC	26
NJ	ATLANTIC CITY	WFPG	27
OH	CINCINNATI	WMOH	28
PA	ERIE	WPSE	31
PA	PITTSBURGH	WJPA	33
PA	STATE COLL.	WMAJ	34
VA	RICHMOND	WCLM	36
WI	WAUSAU	WDLB	38
WV	PARKERSBURG	WLTP	40

	1460		
OH	COLUMBUS	WBNS	29
PA	HARRISBURG	WWKL	31

	1470		
IL	PEORIA	WMBD	20
MI	FLINT	WFNT	25
MI	KALAMAZOO	WQSN	26
OH	TOLEDO	WLQR	30
PA	ALLENTOWN	WKAP	31
WI	MILWAUKEE	WBKV	38

	1480		
IN	TERRE HAUTE	WTHI	22
MI	GRAND RAPIDS	WGVU	26
OH	CANTON	WHBC	28
OH	CINCINNATI	WCIN	28
PA	PHILADELPHIA	WDAS	33
VA	ROANOKE	WTOY	36
WI	MADISON	WTDY	38

	1490		
IL	DANVILLE	WDAN	20
IN	SOUTH BEND	WNDU	22
KY	OWENSBORO	WOMI	23
MD	HAGERSTOWN	WARK	24
MI	SAGINAW	WMPX	27
OH	CLEVELAND	WJMO	29
PA	JOHNSTOWN	WNTJ	32
PA	LANCASTER	WLPA	32
PA	SCRANTON	WAZL	34
WI	APPLETON	WOSH	37
WI	LA CROSSE	WLFN	37
WV	CHARLESTON	WSWW	39
WV	MORGANTOWN	WTCS	39
WV	PARKERSBURG	WMOA	40

	1500		
DC	WASH. D.C.	WTOP	18
IN	INDIANAPOLIS	WBRI	22
MI	BATTLE CREEK	WOLY	25

	1510		
PA	ALLENTOWN	WRNJ	31
WI	MILWAUKEE	WAUK	38

	1520		
OH	TOLEDO	WDMN	30

	1530		
OH	CINCINNATI	WSAI	28

	1540		

		WACA	18
DC	WASH. D.C.	WACA	18
PA	PHILADELPHIA	WNWR	33

	1570		
IN	FT. WAYNE	WGLL	21
WI	APPLETON	WRJQ	37
WI	LA CROSSE	WKBH	37

	1580		
DC	WASH. D.C.	WPGC	18
IN	SOUTH BEND	WHLY	22
OH	COLUMBUS	WVKO	29

	1590		
IN	INDIANAPOLIS	WNTS	22
MD	HAGERSTOWN	WCBG	24
OH	AKRON	WAKR	28

	1600		
MI	ANN ARBOR	WAAM	25
PA	ALLENTOWN	WHOL	31

B

49

C

TENNESSEE

CHATTANOOGA	65
JACKSON	65
JOHNSON CITY-KIN	66
KNOXVILLE	66
MEMPHIS	66
NASHVILLE	67

MISSISSIPPI

BILOXI-GULFPORT	60
JACKSON	60
HATTIESBURG	61
TUPELO	61

NORTH CAROLINA

ASHEVILLE	61
CHARLOTTE	61
FAYETTEVILLE	62
GREENSBORO-WINS	62
GREENVILLE-NEW B	62
RALEIGH-DURHAM	63
WILMINGTON	63

SOUTH CAROLINA

CHARLESTON	64
COLUMBIA	64
FLORENCE	64
GREENVILLE-SPART	64
MYRTLE BEACH	65

GEORGIA

ALBANY	58
ATLANTA	59
AUGUSTA	59
COLUMBUS	60
MACON	60
SAVANNAH	60

ALABAMA

BIRMINGHAM	52
DOTHAN	52
HUNTSVILLE	52
MOBILE	53
MONTGOMERY	53
TUSCALOOSA	53

FLORIDA

DAYTONA BEACH	53
FT. MYERS	53
FT. PIERCE-STUART	54
FT. WALTON BEACH	54
GAINESVILLE	54
JACKSONVILLE	54
KEY WEST	55
LAKELAND-WINTER	55
MELBOURNE	55
MIAMI-FT. LAUDER	56
ORLANDO	56
PANAMA CITY	57
PENSACOLA	57
SARASOTA	57
TALLAHASSEE	57
TAMPA-ST. PETERSB	58
W. PALM BEACH	58

FREQ	STATION	POWER	FORMAT	
94.5	WYSF	|||||||	LGHT AC	
96.5	WMJJ	|||||||	AC	
92.5	WZJT	|||	TOP 40	
99.5	WZRR	|||||||	CLS RK	
107.7	WRAX	|||||||	DVRS	
106.9	WODL	|||||||	OLDS	
95.3	WFFN	|||	CTRY	
97.7	WKLD	|||	CTRY	
102.5	WOWC	|||||||	CTRY	
104.7	WZZK	|||||||	CTRY	
98.7	WBHK	|||||||	BLK, R&B	
AM 1400	WJLD	I	BLK, R&B	
AM 610	WEZN	|||||||	BB, NOS	
90.3	WBHM	|||||||	CLS, NPR, PRI	P
91.1	WVSU	I	JZZ	C
AM 1320	WAGG	|||||	GOSP	
93.7	WDJC	|||||||	RLG	
AM 850	WYDE	|||||||	KIDS	
105.9	WENN	|||||||	URBAN	
AM 900	WATV	|||	URBAN	
AM 960	WERC	|||||||	NWS, TLK	
AM 1070	WAPI	|||||||	NWS, TLK	

		DOTHAN, AL		
99.7	WOOF	|||||||	AC	
106.7	WKMX	|||||||	TOP 40	
102.5	WESP	|||	CLS RK	
100.5	WXUS	|||	OLDS	
95.5	WTVY	|||||||	CTRY	
96.9	WDJR	|||||||	CTRY	
101.1	WZTZ	|||	CTRY	
103.9	WQLS	|||	CTRY	
88.7	WRWA	|||||	CLS, NPR, PRI	P
94.3	WIZB	|||	RLG AC	
93.7	WRJM	|||||||	EZ	
92.1	WJJN	|||	URBAN	
101.3	WAGF	|||	URBAN	
AM 560	WOOF	|||||||	SPRTS	
AM 1450	WWNT	I	NWS, TLK	

		HUNTSVILLE, AL		
95.1	WNDA	|||	AC	
99.1	WAHR	|||||||	AC	
104.3	WZYP	|||||||	TOP 40	
106.1	WTAK	|||	CLS RK	
92.5	WWXQ	|||	OLDS	
94.1	WXQW	|||	OLDS	
102.1	WDRM	|||||||	CTRY	
89.3	WLRH	|||||||	CLS, NPR, PRI	P
90.9	WJAB	|||||	JZZ, NPR	C
AM 1000	WDJL	|||||||	GOSP	
AM 1550	WLOR	|||||||	GOSP	
90.1	WOCG	|||	RLG	C
96.9	WRSA	|||||||	EZ	
AM 730	WUMP	|||||	SPRTS	
AM 770	WVNN	|||||||	NWS, TLK	
AM 1230	WBHP	I	NWS	
AM 1450	WTKI	I	NWS, TLK	
AM 1600	WEUP	|||||	URBAN	

		MOBILE, AL		
92.1	WZEW	|||	AC	
99.9	WMXC	|||||||	AC	
97.5	WABB	|||||||	TOP 40	

AL

AL

	FREQ	STATION	POWER	FORMAT	
	96.1	WRKH	IIIIIIII	CLS RK	
	106.5	WAVH	IIIII	OLDS	
	94.9	WKSJ	IIIIIIII	CTRY	
AM	1310	WHEP	I	BB, NOS	
	91.3	WHIL	III	CLS, NPR, PRI	P
AM	900	WGOK	III	GOSP	
	102.1	WHXT	III	GOSP	
AM	1360	WMOB	IIIII	RLG	
AM	1410	WLVV	IIIII	GOSP	
	92.9	WBLX	IIIIIIII	URBAN	
	98.3	WDLT	IIIIIIII	URBAN	
	105.5	WNSP	III	SPRTS	
AM	660	WDLT	IIIIIIII	URBAN	
AM	710	WNTM	IIIII	NWS, TLK	
AM	1270	WKSJ	IIIII	SPRTS	
AM	1480	WABB	IIIII	NWS, TLK	

MONTGOMERY, AL

	FREQ	STATION	POWER	FORMAT	
	96.1	WRWO	III	AC	
	103.3	WMXS	IIIIIIII	AC	
	98.9	WBAM	IIIIIIII	TOP 40	
	95.1	WXFX	IIIII	CLS RK	
	92.3	WLWI	IIIIIIII	CTRY	
	101.9	WJCC	IIIIIIII	CTRY	
AM	1440	WHHY	IIIII	CTRY	
	90.7	WVAS	IIIIIIII	JZZ, NPR	C
AM	1600	WXVI	IIIII	GOSP	
AM	800	WMGY	III	GOSP	
	97.1	WMCZ	III	URBAN	
AM	740	WMSP	IIIIIIII	SPRTS	
AM	1170	WACV	IIIII	NWS, TLK	

TUSCALOOSA, AL

	FREQ	STATION	POWER	FORMAT	
	100.7	WLXY	III	70's RK	
	105.5	WRTR	III	CLS RK	
	90.7	WVUA	I	AC	C
	91.5	WUAL	IIIIIIII	CLS, NPR, PRI	P
AM	1280	WWPG	IIIII	GOSP	
AM	1420	WACT	IIIII	GOSP	
	92.9	WTUG	IIIIIIII	URBAN	
	95.7	WBHJ	IIIIIIII	URBAN	
AM	1230	WTBC	I	NWS, TLK	

DAYTONA BEACH, FL

	FREQ	STATION	POWER	FORMAT	
	103.3	WVYB	III	TOP 40	
	93.1	WKRO	III	DVRS	
	95.7	WHOG	III	CLS RK	
	105.9	WOCL	IIIIIIII	OLDS	
AM	1230	WSBB	I	BB, NOS	
AM	1340	WROD	I	BB, NOS	
AM	1150	WNDB	I	NWS, TLK	
AM	1380	WELE	IIIII	NWS, TLK	
AM	1490	WXVQ	I	NWS, TLK	

FT. MYERS, FL

FREQ	STATION	POWER	FORMAT
96.9	WINK	IIIIIIII	AC
103.9	WXKB	IIIIIIII	TOP 40
96.1	WRXK	IIIIIIII	CLS RK
99.3	WJBX	IIIII	DVRS
95.3	WOLZ	IIIII	OLDS
101.9	WWGR	IIIIIIII	CTRY
105.5	WQNU	IIIIIIII	CTRY

	FREQ	STATION	POWER	FORMAT	
	107.1	WCKT	IIIIIII	CTRY	
AM	1410	WMYR	IIIII	CTRY	
AM	1440	WWCL	IIIII	SPAN	
	98.5	WDRR	III	JZZ	
	106.3	WJST	III	BB, NOS	
	90.1	WGCU	IIIIIIIII	CLS, NPR, PRI	P
AM	770	WWCN	IIIIIIIII	SPRTS	
AM	1200	WTLQ	IIIII	NWS, TLK	
AM	1240	WINK	I	NWS, TLK	

FT. PIERCE, FL

	FREQ	STATION	POWER	FORMAT	
	93.7	WGYL	IIIII	AC	
	94.7	WBBE	IIIIIIII	AC	
	102.3	WMBX	III	HOT AC	
	92.7	WZZR	IIIII	ROCK	
	103.7	WQOL	IIIIIIIII	OLDS	
AM	1450	WSTU	I	OLDS	
	99.7	WPAW	III	CTRY	
	101.7	WAVW	III	CTRY	
	97.1	WOSN	III	BB, NOS	
AM	1400	WIRA	I	BB, NOS	
AM	1490	WTTB	I	BB, NOS	
	88.9	WQCS	IIIII	CLS, NPR, PRI	P
	104.7	WFLM	III	URBAN	
AM	1330	WJNX	IIIII	NWS, TLK	
AM	1370	WAXE	I	NWS, TLK	

FT. WALT. BEACH, FL

	FREQ	STATION	POWER	FORMAT	
	99.5	WKSM	IIIII	CLS RK	
	92.1	WMMK	III	OLDS	
	104.7	WAAZ	IIIII	CTRY	
	105.5	WYZB	III	CTRY	
AM	1400	WFAV	I	BB, NOS	
AM	1260	WFTW	III	NWS, TLK	
AM	1340	WFSH	I	NWS, TLK	

GAINESVILLE, FL

	FREQ	STATION	POWER	FORMAT	
	105.3	WYKS	III	TOP 40	
	92.5	WNDT	III	ROCK	
	95.5	WNDD	III	ROCK	
	103.7	WRUF	IIIIIIIII	ROCK	
	104.9	WRKG	III	OLDS	
	93.7	WOGK	IIIIIIIII	CTRY	
	100.9	WYGC	III	CTRY	
	102.3	WTRS	IIIII	CTRY	
AM	980	WLUS	IIIIIIIII	BB, NOS	
	89.1	WUFT	IIIIIIIII	CLS, NPR, PRI	P
	92.9	WMFQ	IIIII	EZ	
	97.3	WSKY	IIIII	NWS, TLK	
AM	850	WRUF	IIIIIIIII	NWS, TLK	
AM	1230	WGGG	I	SPRTS	
AM	1290	WTMC	IIIII	NWS	
AM	1430	WWLO	III	NWS, TLK	

JACKSONVILLE, FL

	FREQ	STATION	POWER	FORMAT	
	94.1	WSOS	III	HOT AC	
	96.1	WEJZ	IIIIIIIII	AC	
	102.9	WMXQ	IIIIIIIII	AC	
	95.1	WAPE	IIIII	TOP 40	
	93.3	WPLA	IIIII	DVRS	
	104.5	WFYV	IIIIIIIII	ROCK	
	106.5	WBGB	III	CLS RK	

FL

	FREQ	STATION	POWER	FORMAT	
	96.9	WKQL	IIIIIIII	OLDS	
	99.1	WQIK	IIIIIIII	CTRY	
	105.5	WJQR	III	CTRY	
	107.3	WROO	IIIIIIII	CTRY	
	97.9	WFSJ	III	JZZ	
	88.5	WFCF	I	DVRS	C
	89.9	WJCT	IIIIIIII	CLS, NPR, PRI	P
AM	1400	WZAZ	I	GOSP	
	90.9	WKTZ	IIIII	EZ	P
AM	550	WAYR	IIIIIIII	RLG	P
AM	1280	WSVE	IIIII	GOSP	
	92.1	WJXR	III	NWS, TLK	
	92.7	WJBT	III	URBAN	
	105.7	WXQL	III	URBAN	
AM	600	WBWL	IIIIIIII	SPRTS	
AM	690	WOKV	IIIIIIII	NWS, TLK	
AM	930	WNZS	IIIIIIII	SPRTS	
AM	1240	WFOY	I	NWS, TLK	
AM	1320	WJGR	IIIII	NWS, TLK	
AM	1420	WAOC	III	NWS, TLK	
AM	1460	WZNZ	IIIII	NWS	

KEY WEST, FL

	FREQ	STATION	POWER	FORMAT
	93.5	WKRY	III	AC
	94.3	WGMX	IIIII	AC
	100.3	WCTH	IIIII	AC
	102.1	WKLG	III	AC
	103.1	WFKZ	III	AC
	105.5	WAVK	III	AC
	92.5	WEOW	IIIIIIII	TOP 40
	99.5	WAIL	IIIIIIII	CLS RK
	107.1	WIIS	I	ALT RK
	104.7	WWUS	IIIII	OLDS
	102.5	WPIK	IIIII	CTRY
	106.3	WZMQ	III	SPAN
	107.9	WVMQ	IIIIIIII	SPAN
AM	1500	WKIZ	I	SPAN
	98.7	WCNK	III	JZZ
	90.9	WJIR	I	RLG
AM	1300	WFFG	III	NWS, TLK
AM	1600	WKWF	I	SPRTS

LAKELAND, FL

	FREQ	STATION	POWER	FORMAT
AM	1130	WWBF	III	OLDS
	97.5	WPCV	IIIIIIII	CTRY
AM	1460	WBAR	I	CTRY
AM	1230	WONN	I	BB, NOS
AM	1330	WWAB	I	GOSP
AM	1360	WHNR	IIIII	URBAN
AM	1430	WLKF	IIIII	NWS, TLK

MELBOURNE, FL

	FREQ	STATION	POWER	FORMAT	
	99.3	WLRQ	IIIII	AC	
	107.1	WAOA	IIIII	TOP 40	
	95.1	WBVD	I	CLS RK	
	102.7	WHKR	IIIIIIII	CTRY	
AM	1240	WMMB	I	BB, NOS	
AM	1350	WMMV	I	BB, NOS	
AM	1560	WTMS	IIIII	BB, NOS	
	89.5	WFIT	I	JZZ, NPR	P
AM	1300	WXXU	IIIII	GOSP	
AM	860	WRFB	III	EZ	

	FREQ	STATION	POWER	FORMAT	
AM	1510	WWBC	I	RLG	
AM	920	WMEL	IIIIIIII	NWS, TLK	
AM	1060	WAMT	IIIII	NWS, TLK	

MIAMI, FL

	FREQ	STATION	POWER	FORMAT	
	97.3	WFLC	IIIIIIII	AC	
	101.5	WLYF	IIIIIIII	AC	
	103.5	WPLL	IIIIIIII	AC	
AM	610	WIOD	IIIIIIII	AC	
	96.5	WPOW	IIIIIIII	TOP 40	
	100.7	WHYI	IIIIIIII	TOP 40	
	94.9	WZTA	IIIIIIII	MOD RK	
	105.9	WBGG	IIIIIIII	CLS RK	
	102.7	WMXJ	IIIIIIII	OLDS	
	92.3	WCMQ	IIIII	SPAN	
	95.7	WXDJ	IIIII	SPAN	
	98.3	WRTO	IIIIIIII	SPAN	
	106.7	WRMA	IIIIIIII	SPAN	
	107.5	WAMR	IIIIIIII	SPAN	
AM	670	WWFE	IIIIIIII	SPAN	
AM	710	WAQI	IIIIIIII	SPAN	
AM	830	WACC	III	SPAN	
AM	1140	WQBA	IIIIIIII	SPAN	
AM	1210	WNMA	IIIII	SPAN	
AM	1260	WSUA	IIIII	SPAN	
AM	1360	WKAT	IIIII	SPAN	
	93.1	WTMI	IIIIIIII	CLS	
	93.9	WLVE	IIIIIIII	JZZ	
	88.9	WDNA	III	JZZ, CLS, PRI	P
	90.5	WVUM	I	AC	C
	91.3	WLRN	IIIIIIII	DVRS, NPR, PRI	P
AM	1490	WMBM	I	GOSP	
AM	980	WHSR	IIIIIIII	ETH	
AM	1080	WVCG	IIIIIIII	ETH	
AM	1170	WAVS	IIIII	ETH	
AM	1320	WLQY	IIIII	ETH	
AM	1520	WEXY	III	RLG	
	99.1	WEDR	IIIIIIII	URBAN	
	105.1	WHQT	IIIIIIII	URBAN	
AM	560	WQAM	IIIIIIII	SPRTS	
AM	790	WAXY	IIIIIIII	NWS, TLK	
AM	940	WINZ	IIIIIIII	NWS, TLK	
AM	1400	WFTL	I	NWS, TLK	

ORLANDO, FL

	FREQ	STATION	POWER	FORMAT	
	98.9	WMMO	III	LGHT AC	
	100.3	WSHE	IIIIIIII	AC	
	105.1	WOMX	IIIIIIII	AC	
	96.5	WHTQ	IIIIIIII	CLS RK	
	92.3	WWKA	IIIIIIII	CTRY	
AM	1030	WONQ	IIIII	SPAN	
AM	1080	WFIV	IIIII	SPAN	
AM	1270	WRLZ	IIIII	SPAN	
AM	1440	WPRD	IIIII	SPAN	
	103.1	WLOQ	IIIII	JZZ	
AM	990	WHOO	IIIIIIII	BB, NOS	
	89.9	WUCF	III	JZZ, NPR	C
	90.7	WMFE	IIIIIIII	CLS, NPR, PRI	P
	91.5	WPRK	I	DVRS	C
AM	1600	WOKB	IIIII	GOSP	
	95.3	WTLN	III	RLG	
AM	540	WQTM	IIIIIIII	SPRTS	

FREQ	STATION	POWER	FORMAT	
AM 580	WDBO	IIIIIII	NWS, TLK	
AM 740	WWNZ	IIIIIII	NWS. TLK	

PANAMA CITY, FL

FREQ	STATION	POWER	FORMAT	
98.5	WFSY	IIIIIII	AC	
99.3	WSHF	IIIIIII	AC	
105.9	WILN	IIIII	HOT AC	
103.5	WMXP	IIIIIII	TOP 40	
95.9	WRBA	IIIII	CLS RK	
107.9	WDRK	IIIII	ROCK	
92.5	WPAP	IIIIIII	CTRY	
105.1	WAKT	IIIII	CTRY	
89.1	WFSW	IIIII	NWS, TLK, NPR, PRI	P
90.7	WKGC	IIIII	CLS, NPR, PRI	C
100.1	WPCF	III	RLG AC	
AM 1480	WKGC	I	NWS, NPR	C
AM 590	WDIZ	IIIIIII	EZ	
101.1	WYOO	III	NWS, TLK	
AM 1430	WLTG	IIIII	NWS, TLK	

PENSACOLA, FL

FREQ	STATION	POWER	FORMAT	
94.1	WMEZ	IIIIIII	LGHT AC	
100.7	WWRO	IIIIIII	CLS RK	
101.5	WTKX	IIIII	MOD RK	
107.3	WYCL	IIIIIII	OLDS	
AM 1450	WBSR	I	OLDS	
102.7	WXBM	IIIIIII	CTRY	
AM 980	WRNE	IIIII	BLK, R&B	
88.1	WUWF	IIIIIII	NWS, TLK, NPR, PRI	C
AM 610	WVTJ	IIIII	RLG	
AM 1330	WEBY	IIIII	NWS, TLK	
AM 1370	WCOA	IIIII	NWS, TLK	

SARASOTA, FL

FREQ	STATION	POWER	FORMAT	
98.7	WLLD	IIIII	TOP 40	
107.9	WYNF	IIIII	CLS RK	
106.5	WSRZ	III	OLDS	
AM 1490	WWPR	I	OLDS	
92.1	WCTQ	III	CTRY	
88.1	WJIS	IIIII	RLG	P
103.5	WDUV	IIIIIII	EZ	
AM 1420	WBRD	III	GOSP	
AM 930	WKXY	IIIIIII	NWS, TLK	
AM 1220	WQSA	I	SPRTS	
AM 1320	WAMR	IIIII	SPRTS	
AM 1530	WENG	I	NWS, TLK	

TALLAHASSEE, FL

FREQ	STATION	POWER	FORMAT	
94.1	WAKU	III	HOT AC	
98.9	WBZE	IIIII	AC	
104.9	WFLV	IIIIIII	AC	
106.1	WWLD	III	TOP 40	
99.9	WWFO	IIIIIII	CLS RK	
101.5	WXSR	IIIII	MOD RK	
104.1	WGLF	IIIIIII	CLS RK	
94.9	WTNT	IIIIIII	CTRY	
103.1	WAIB	IIIII	CTRY	
88.9	WFSU	IIIIIII	NWS, NPR, PRI	P
89.7	WVFS	I	AC	C
90.5	WAMF	I	DVRS	C
91.5	WFSQ	IIIIIII	CLS, NPR, PRI	P
96.1	WHBX	III	URBAN	

	FREQ	STATION	POWER	FORMAT	
AM	1270	WNLS	IIIII	SPRTS	
AM	1450	WTAL	I	NWS. TLK	

TAMPA, FL

	FREQ	STATION	POWER	FORMAT	
	95.7	WSSR	IIIIIIII	AC	
	97.1	WLVU	III	AC	
	100.7	WAKS	IIIIIIII	AC	
	93.3	WFLZ	IIIIIIII	TOP 40	
	97.9	WXTB	IIIIIIII	ROCK	
	105.5	WTBT	III	CLS RK	
	107.3	WCOF	IIIIIIII	70's RK	
AM	910	WFNS	IIIIIIII	70's RK	
	92.5	WYUU	IIIII	OLDS	
	99.5	WQYK	IIIIIIII	CTRY	
	104.7	WRBQ	IIIIIIII	CTRY	
AM	760	WBDN	IIIIIIII	SPAN	
AM	1550	WAMA	IIIII	SPAN	
	94.1	WSJT	IIIIIIII	JZZ	
	96.1	WGUL	III	BB, NOS	
AM	860	WGUL	IIIIIIII	BB, NOS	
	88.5	WMNF	IIIIIIII	DVRS, NPR	P
	89.7	WUSF	IIIIIIII	CLS, NPR, PRI	P
	94.9	WWRM	IIIIIIII	EZ	
AM	1470	WRTB	IIIII	ETH	
AM	1380	WRBQ	IIIII	URBAN	
AM	1590	WRXB	IIIII	URBAN	
AM	1150	WTMP	IIIII	URBAN	
AM	570	WHNZ	IIIIIIII	NWS, TLK	
AM	620	WSUN	IIIIIIII	NWS	
AM	820	WZTM	IIIIIIII	SPRTS	
AM	970	WFLA	IIIIIIII	NWS, TLK	
AM	1010	WQYK	IIIIIIII	NWS, TLK	
AM	1250	WDAE	IIIII	SPRTS	

W. PALM BEACH, FL

	FREQ	STATION	POWER	FORMAT	
	92.1	WRLX	III	LGHT AC	
	97.9	WRMF	IIIIIIII	AC	
	104.3	WEAT	IIIIIIII	AC	
	105.5	WWLV	IIIII	OLDS	
	107.9	WIRK	IIIII	CTRY	
AM	1380	WLVS	I	SPAN	
	99.5	WJBW	III	BB, NOS	
AM	1230	WJNA	I	BB, NOS	
AM	1420	WDBF	IIIII	BB, NOS	
	90.7	WXEL	III	CLS, NPR, PRI	P
AM	1600	WPOM	IIIII	GOSP	
AM	640	WLVJ	IIIIIIII	RLG	
AM	740	WSBR	IIIIIIII	BUS NWS	
AM	850	WDJA	IIIIIIII	BUS NWS	
AM	1040	WJNO	IIIII	NWS, TLK	
AM	1290	WBZT	IIIII	NWS, TLK	
AM	1340	WPBR	I	NWS, TLK	

ALBANY, GA

	FREQ	STATION	POWER	FORMAT	
	103.5	WJAD	III	CLS RK	
	101.7	WKAK	III	CTRY	
	91.7	WUNV	I	CLS, NPR, PRI	P
AM	960	WJYZ	IIIIIIII	GOSP	
	104.5	WGPC	IIIIIIII	EZ	
	106.5	WZIQ	III	RLG	
AM	1450	WGPC	I	EZ	
	96.3	WJIZ	IIIII	URBAN	

	FREQ	STATION	POWER	FORMAT	
AM	1590	WALG	I	NWS, TLK	

ATLANTA, GA

	FREQ	STATION	POWER	FORMAT	
	94.9	WPCH	IIIIIII	AC	
	98.5	WSB	IIIIIII	AC	
	94.1	WSTR	IIIIIII	TOP 40	
	92.9	WZGC	IIIIIII	CLS RK	
	96.1	WKLS	IIIIIII	ROCK	
	99.7	WNNX	IIIIIII	DVRS	
	101.5	WKHX	IIIIIII	CTRY	
AM	1550	WAZX	IIIIIII	SPAN	
	88.5	WRAS	IIIII	DVRS, PRI	C
	89.3	WRFG	IIIII	DVRS	P
	90.1	WABE	IIIIIII	CLS, NPR, PRI	P
	90.7	WWGC	III	DVRS, NPR, PRI	C
	91.1	WREK	III	DVRS	C
	91.9	WCLK	III	JZZ, NPR	C
AM	1340	WALR	I	GOSP	
AM	1480	WYZE	IIIII	GOSP	
AM	970	WNIV	IIIIIII	RLG	
AM	1380	WAOK	IIIII	GOSP	
	97.5	WHTA	III	URBAN	
	103.3	WVEE	IIIIIII	URBAN	
	105.7	WGST	IIIII	NWS, TLK	
AM	640	WGST	IIIIIII	NWS, TLK	
AM	680	WCNN	IIIIIII	NWS	
AM	750	WSB	IIIIIII	NWS, TLK	
AM	790	WQXI	IIIIIII	SPRTS	

AUGUSTA, GA

	FREQ	STATION	POWER	FORMAT	
	98.3	WSLT	III	AC	
	104.3	WBBQ	IIIIIII	AC	
AM	1340	WBBQ	I	AC	
	105.7	WZNY	IIIIIII	TOP 40	
	95.1	WCHZ	IIIII	70's RK	
	96.3	WRXR	IIIIIII	MOD RK	
	102.3	WEKL	III	CLS RK	
	93.9	WGOR	III	OLDS	
	99.5	WKXC	IIIIIII	CTRY	
	101.7	WTHO	III	CTRY	
	107.7	WUUS	IIIII	CTRY	
	92.3	WAEG	III	JZZ	
	102.7	WAJY	III	BB, NOS	
	90.7	WACG	III	CLS, NPR, PRI	P
AM	1550	WTHB	IIIII	GOSP	
	103.1	WFXA	III	URBAN	
AM	580	WGAC	IIIIIII	NWS, TLK	
AM	1230	WKIM	I	URBAN	
AM	1480	WRDW	IIIII	SPRTS	

COLUMBUS, GA

	FREQ	STATION	POWER	FORMAT	
	100.1	WGSY	III	AC	
	107.3	WCGQ	IIIIIII	AC	
	102.9	WVRK	IIIIIII	ROCK	
	91.7	WTJB	III	CLS, NPR, PRI	P
AM	1340	WOKS	I	GOSP	
AM	1460	WPNX	IIIII	GOSP	
AM	1580	WEAM	III	RLG	
	98.3	WAGH	III	URBAN	
	104.9	WFXE	III	URBAN	
AM	540	WDAK	IIIIIII	SPRTS	
AM	1420	WRCG	IIIII	NWS, TLK	

	FREQ	STATION	POWER	FORMAT	
	100.9	WPGA	III	HOT AC	
	101.7	WRBV	III	AC	
	107.9	WPEZ	IIIIIII	LGHT AC	
	92.3	WPPG	III	CLS RK	
	106.3	WQBZ	IIIII	ROCK	
	99.1	WAYS	IIIIIIII	OLDS	
AM	980	WPGA	IIIIIIII	OLDS	
	105.3	WDEN	IIIIIIII	CTRY	
	102.5	WLCG	III	GOSP	
AM	1240	WDDO	I	GOSP	
AM	1350	WCOP	IIIII	RLG	
	96.5	WIBD	III	URBAN	
	97.9	WIBB	III	URBAN	
AM	940	WMAC	IIIIIIII	NWS, TLK	

SAVANNAH, GA

	FREQ	STATION	POWER	FORMAT	
	97.3	WAEV	IIIIIIII	AC	
	102.1	WZAT	IIIIIIII	TOP 40	
	95.5	WIXV	IIIIIIII	ROCK	
	105.3	WRHQ	III	ROCK	
	94.1	WCHY	IIIIIIII	CTRY	
	96.5	WJCL	IIIIIIII	CTRY	
	90.3	WHCJ	I	JZZ	C
	91.1	WSVH	IIIIIIII	CLS, NPR, PRI	P
AM	1230	WSOK	I	GOSP	
	93.1	WEAS	IIIIIIII	URBAN	
	103.9	WSGF	III	URBAN	
AM	630	WBMQ	IIIIIIII	NWS, TLK	
AM	900	WEAS	IIIIIIII	SPRTS	

GULFPORT, MS

	FREQ	STATION	POWER	FORMAT	
	93.7	WMJY	IIIIIIII	AC	
	96.7	WLRK	III	LGHT AC	
	107.1	WXYK	III	TOP 40	
	105.9	WXRG	III	CLS RK	
	102.3	WGCM	III	OLDS	
	99.1	WKNN	IIIIIIII	CTRY	
AM	1490	WXBD	I	BB, NOS	
	90.3	WMAH	IIIIIIII	CLS, NPR, PRI	P
	103.1	WOSM	IIIIIIII	GOSP	
AM	1390	WROA	IIIII	EZ	
	94.5	WJZD	III	URBAN	
	104.9	WYOK	IIIII	URBAN	
AM	570	WVMI	IIIIIIII	NWS, TLK	

JACKSON, MS

	FREQ	STATION	POWER	FORMAT	
	96.3	WJDX	IIIIIIII	AC	
	101.7	WYOY	III	TOP 40	
	94.7	WTYX	IIIIIIII	CLS RK	
AM	1400	WKXI	I	OLDS	
	95.5	WKTF	IIIIIIII	CTRY	
	102.9	WMSI	IIIIIIII	CTRY	
	93.9	WVIV	III	BB, NOS	
	88.5	WJSU	I	JZZ, NPR	P
	90.1	WMPR	IIIIIIII	DVRS	P
	91.3	WMPN	IIIIIIII	CLS, NPR, PRI	P
	93.5	WHJT	III	RLG AC	
AM	1300	WOAD	IIIII	GOSP	
AM	1590	WZRX	IIIII	GOSP	
	92.9	WJXN	III	GOSP	

GA
MS

	FREQ	STATION	POWER	FORMAT	
	97.7	WRJH	III	GOSP	
	99.7	WJMI	IIIIIIII	URBAN	
AM	620	WJDS	IIIIIIII	SPRTS	
AM	1180	WJNT	IIIIIIII	NWS, TLK	

HATTIESBURG, MS

	FREQ	STATION	POWER	FORMAT	
	106.3	WMFM	III	LGHT AC	
	100.3	WNSL	IIIIIIII	TOP 40	
	104.5	WXRR	IIIIIIII	CLS RK	
	103.7	WHER	IIIIIIII	OLDS	
	88.5	WUSM	III	CLS	P
	98.1	WMXI	III	RLG AC	
AM	1340	WAML	I	GOSP	
AM	950	WBKH	III	GOSP	
AM	1400	WFOR	I	GOSP	
	92.1	WJMG	III	URBAN	
	102.5	WJKX	IIIIIIII	URBAN	

TUPELO, MS

	FREQ	STATION	POWER	FORMAT	
	93.3	WSYE	IIIIIIII	AC	
	101.9	WFTA	IIIII	AC	
	95.3	WAFM	III	OLDS	
	105.3	WWKZ	III	OLDS	
	98.5	WZLQ	IIIII	CTRY	
	106.7	WWZD	IIIII	CTRY	
AM	1330	WFTO	IIIII	CTRY	
	103.5	WRBO	IIIIIIII	BLK, R&B	
	89.5	WMAE	IIIIIIII	CLS, NPR, PRI	P
	96.7	WSEL	III	GOSP	
	99.3	WBIP	III	GOSP	
AM	940	WCPC	IIIIIIII	GOSP	
AM	1470	WNAU	I	GOSP	
	92.5	WESE	III	URBAN	
AM	580	WELO	IIIII	NWS, TLK	

ASHEVILLE, NC

	FREQ	STATION	POWER	FORMAT	
	96.5	WZLS	III	ROCK	
	99.9	WKSF	IIIIIIII	CTRY	
AM	570	WWNC	IIIIIIII	CTRY	
AM	1310	WISE	IIIII	BB, NOS	
	88.1	WCQS	III	NWS, TLK, NPR, PRI	P
AM	1230	WSKY	I	RLG AC	
AM	1460	WHBK	IIIII	GOSP	

CHARLOTTE, NC

	FREQ	STATION	POWER	FORMAT	
	101.9	WBAV	IIIIIIII	AC	
	107.9	WLNK	IIIIIIII	AC	
	95.1	WNKS	IIIIIIII	TOP 40	
	99.7	WRFX	IIIIIIII	ROCK	
	104.7	WSSS	IIIIIIII	70's RK	
	106.5	WEND	IIIIIIII	DVRS	
	103.7	WSOC	IIIIIIII	CTRY	
AM	1050	WLON	I	CTRY	
	92.7	WCCJ	III	JZZ	
	106.1	WNMX	III	BB, NOS	
	89.9	WDAV	IIIIIIII	CLS, NPR, PRI	P
	90.7	WFAE	IIIIIIII	NWS, TLK, NPR, PRI	P
	91.7	WSGE	I	DVRS	C
AM	1600	WGIV	I	GOSP	
AM	1240	WHVN	I	RLG	
AM	1270	WCGC	IIIII	RLG	
AM	1370	WLTC	IIIII	GOSP	

MS
NC

NC

	FREQ	STATION	POWER	FORMAT	
AM	1480	WGFY	|||||	KIDS	
	97.9	WPEG	|||||	URBAN	
AM	610	WFNZ	||||||||	SPRTS	
AM	1110	WBT	|||||||	NWS, TLK	
AM	1340	WRHI	I	NWS, TLK	
AM	1490	WSTP	I	NWS, TLK	

FAYETTEVILLE, NC

	FREQ	STATION	POWER	FORMAT	
	98.1	WQSM	||||||||	HOT AC	
	95.7	WKML	||||||||	CTRY	
AM	1490	WAZZ	I	BB, NOS	
	91.9	WFSS	|||||	JZZ, NPR, PRI	P
AM	1600	WIDU	|||||	GOSP	
	100.9	WSTS	||||||||	GOSP	
AM	1160	WYRU	|||||	GOSP	
	107.7	WUKS	|||	URBAN	
AM	640	WFNC	|||||||	NWS, TLK	
AM	1230	WFAI	I	NWS, TLK	

GREENS. / WIN. SAL., NC

	FREQ	STATION	POWER	FORMAT	
	93.9	WRSN	||||||||	LGHT AC	
	99.5	WMAG	||||||||	AC	
	107.5	WKZL	|||||	HOT AC	
	92.3	WKRR	||||||||	CLS RK	
	98.7	WKSI	||||||||	DVRS	
	93.1	WMQX	||||||||	OLDS	
	100.3	WHSL	||||||||	CTRY	
	101.1	WKXU	||||||||	CTRY	
	104.1	WTQR	||||||||	CTRY	
AM	1070	WGOS	I	CTRY	
AM	1260	WKXR	|||||	CTRY	
AM	980	WAAA	|||	BLK, R&B	
	98.3	WIST	|||	BB, NOS	
	88.5	WFDD	|||	CLS, NPR, PRI	P
	89.3	WSOE	I	DVRS	C
	90.1	WNAA	|||	URBAN	C
	90.5	WWCU	I	TOP 40	C
	90.5	WSNC	|||	DVRS	P
	90.9	WQFS	I	DVRS	C
	94.1	WWGL	||||||||	RLG AC	
AM	1340	WPOL	I	GOSP	
AM	1400	WKEW	I	KIDS	
	97.1	WQMG	||||||||	URBAN	
AM	600	WSJS	||||||||	NWS, TLK	
AM	1230	WMFR	I	NWS, TLK	
AM	1320	WTCK	||||	NWS, TLK	
AM	1440	WLXN	|||||	NWS, TLK	

GREEN. / NEW BERN, NC

	FREQ	STATION	POWER	FORMAT	
	94.3	WGPM	|||	HOT AC	
	103.3	WMGV	||||||||	AC	
	92.3	WQSL	|||	TOP 40	
	96.3	WRHT	||||||||	TOP 40	
	93.3	WERO	||||||||	CLS RK	
	99.5	WXNR	||||||||	DVRS	
	105.5	WXQR	|||||	CLS RK	
	106.5	WSFL	||||||||	CLS RK	
	98.7	WKOO	||||||||	OLDS	
	101.1	WKJA	||||||||	OLDS	
	107.9	WNCT	||||||||	OLDS	
	94.1	WNBR	|||||	CTRY	
	95.1	WRNS	||||||||	CTRY	

	FREQ	STATION	POWER	FORMAT	
	97.7	WZBR	I	CTRY	
AM	1070	WNCT	IIIII	BLK, R&B	
	105.1	WANG	III	BB, NOS	
AM	1450	WNOS	I	BB, NOS	
	89.3	WTEB	IIIIIIII	NWS, NPR, PRI	P
	90.3	WKNS	III	NWS, NPR, PRI	P
	91.3	WZMB	I	AC	C
	102.9	WELS	III	GOSP	
AM	1340	WOOW	I	GOSP	
AM	1010	WELS	I	GOSP	
AM	1250	WGHB	IIIII	GOSP	
AM	1290	WJCV	I	GOSP	
	98.3	WCZI	III	NWS, TLK	
	101.9	WIKS	IIIIIIII	URBAN	
	107.3	WTKF	III	NWS, TLK	
AM	930	WDLX	IIIIIIII	NWS, TLK	
AM	1240	WJNC	I	NWS, TLK	

RALEIGH-DURHAM, NC

	FREQ	STATION	POWER	FORMAT	
	101.5	WRAL	IIIIIIII	AC	
AM	620	WDNC	IIIIIIII	AC	
	96.1	WBBB	IIIIIIII	TOP 40	
	105.1	WDCG	IIIIIIII	TOP 40	
AM	1360	WCHL	IIIII	OLDS	
	94.7	WQDR	IIIIIIII	CTRY	
	102.5	WHLQ	III	CTRY	
AM	540	WETC	IIIIIIII	CTRY	
AM	1090	WBZB	I	CTRY	
AM	1270	WMPM	IIIII	CTRY	
	88.1	WKNC	III	ROCK	C
	88.7	WXDU	III	DVRS	C
	88.9	WSHA	III	JZZ	P
	89.3	WXYC	III	AC	C
	90.7	WNCU	IIIII	JZZ, NPR, PRI	C
	91.5	WUNC	IIIIIIII	CLS, NPR, PRI	P
	103.9	WNNL	III	GOSP	
AM	1240	WPJL	I	RLG AC	
AM	1410	WSRC	IIIII	GOSP	
AM	1490	WDUR	I	GOSP	
	107.1	WFXC	III	URBAN	
AM	680	WPTF	IIIIIIII	NWS, TLK	
AM	850	WRBZ	IIIIIIII	NWS, TLK	

WILMINGTON, NC

	FREQ	STATION	POWER	FORMAT	
	102.7	WGNI	IIIIIIII	AC	
	104.5	WRQR	III	ROCK	
	107.5	WSFM	IIIIIIII	MOD RK	
	103.7	WLTT	III	OLDS	
	106.3	WCCA	III	OLDS	
	101.3	WWQQ	IIIIIIII	CTRY	
	91.3	WHQR	IIIII	CLS, NPR, PRI	P
AM	1490	WWIL	I	GOSP	
	97.3	WMNX	IIIIIIII	URBAN	
AM	630	WMFD	IIIII	NWS, TLK	
AM	980	WAAV	IIIIIIII	NWS, TLK	

CHARLESTON, SC

	FREQ	STATION	POWER	FORMAT	
	95.1	WSSX	IIIII	HOT AC	
	96.9	WSUY	IIIIIIII	AC	
	100.5	WLLC	IIIII	AC	
	105.3	WNST	III	HOT AC	
	96.1	WAVF	IIIIIIII	MOD RK	
	98.1	WYBB	IIIII	CLS RK	
	104.5	WRFQ	IIIIIIII	CLS RK	

63

FREQ	STATION	POWER	FORMAT	
98.9	WWBZ	IIIII	OLDS	
102.5	WXLY	IIIIIIII	OLDS	
103.5	WEZL	IIIIIIII	CTRY	
107.5	WNKT	IIIIIIII	CTRY	
AM 730	WPAL	IIIII	BLK, R&B	
94.3	WSSP	III	BB, NOS	
89.3	WSCI	IIIIIIII	CLS, NPR, PRI	P
106.1	WTUA	III	GOSP	
AM 1390	WXTC	IIIII	GOSP	
93.3	WWWZ	IIIII	URBAN	
101.7	WMGL	III	URBAN	
AM 910	WTMZ	III	NWS	
AM 1250	WTMA	IIIII	NWS, TLK	
AM 1340	WQSC	I	SPRTS	
AM 1450	WQNT	I	NWS, TLK	

COLUMBIA, SC

104.7	WNOK	IIIIIIII	TOP 40	
93.5	WARQ	III	MOD RK	
102.3	WMFX	III	CLS RK	
103.1	WOMG	III	OLDS	
96.7	WLTY	III	CTRY	
97.5	WCOS	IIIIIIII	CTRY	
98.5	WLXC	III	BLK, R&B	
100.1	WSCQ	III	BB, NOS	
91.3	WLTR	IIIIIIII	CLS, NPR, PRI	P
95.3	WFMV	III	RLG AC	
AM 620	WTGH	IIIII	RLG	
AM 1470	WQXL	IIIII	RLG	
AM 560	WVOC	IIIIIIII	NWS, TLK	
AM 1230	WOIC	I	URBAN	
AM 1320	WISW	IIIII	NWS	
AM 1400	WCOS	I	NWS, TLK	

FLORENCE, SC

102.1	WMXT	IIIII	AC	
98.5	WHSC	III	CLS RK	
100.1	WWFN	III	OLDS	
AM 1230	WOLS	I	OLDS	
105.5	WDAR	III	CTRY	
AM 1450	WHSC	I	CTRY	
102.9	WSQN	III	BLK, R&B	
AM 540	WYNN	IIIII	GOSP	
106.3	WYNN	III	URBAN	
AM 970	WJMX	IIIIIIII	NWS, TLK	

GREEN. / SPARTBRG., SC

98.9	WSPA	IIIIIIII	LGHT AC	
93.7	WFBC	IIIIIIII	TOP 40	
101.1	WROQ	IIIIIIII	CLS RK	
103.3	WOLT	III	OLDS	
103.9	WOLI	III	OLDS	
92.5	WESC	IIIIIIII	CTRY	
103.1	WRIX	III	CTRY	
AM 1400	WKDY	I	CTRY	
AM 1580	WDAB	IIIII	SPAN	
88.1	WSBF	I	DVRS	C
88.5	WEPC	IIIII	RLG	C
94.5	WMUU	IIIIIIII	EZ	
AM 1020	WRIX	IIIII	GOSP	
AM 1300	WCKI	I	GOSP	
AM 1440	WGVL	IIIII	GOSP	

	FREQ	STATION	POWER	FORMAT	
	104.9	WCCP	III	SPRTS	
	107.3	WJMZ	IIIIIIIII	URBAN	
AM	910	WORD	IIIII	NWS, TLK	
AM	950	WSPA	IIIIIIII	NWS, TLK	
AM	1070	WHYZ	IIIIIIIII	NWS, TLK	
AM	1330	WYRD	IIIII	NWS, TLK	

MYRTLE BEACH, SC

	FREQ	STATION	POWER	FORMAT	
	99.5	WMYB	III	AC	
	105.9	WNMB	III	AC	
	93.9	WJXY	III	TOP 40	
	97.7	WWXM	IIIII	TOP 40	
	101.7	WKZQ	III	ROCK	
	104.1	WYAV	IIIIIIIII	CLS RK	
	106.5	WSYN	IIIII	OLDS	
	100.7	WGTN	III	CTRY	
	103.1	WYAK	III	CTRY	
	107.9	WGTR	IIIII	CTRY	
	94.9	WVCO	III	JZZ	
AM	1050	WJXY	IIIII	BB, NOS	
AM	1470	WLMC	I	GOSP	
	92.1	WJYR	IIIII	EZ	
AM	1240	WLSC	I	EZ	
AM	1330	WPJS	IIIII	GOSP	
	94.5	WRNN	III	NWS, TLK	
	98.5	WDAI	III	URBAN	

CHATTANOOGA, TN

	FREQ	STATION	POWER	FORMAT	
	92.3	WDEF	IIIIIIII	AC	
	105.5	WLMX	III	AC	
AM	980	WLMX	III	AC	
	97.3	WKXJ	IIIIIIIII	TOP 40	
	101.9	WSGC	III	TOP 40	
	96.5	WDOD	IIIIIIII	DVRS	
	106.5	WSKZ	IIIIIIIII	CLS RK	
	107.9	WOGT	IIIII	OLDS	
	98.1	WXKT	III	CTRY	
AM	1190	WSDQ	IIIII	CTRY	
	93.7	WMPZ	III	JZZ	
AM	1310	WDOD	IIIII	BB, NOS	
	88.1	WUTC	IIIII	JZZ, NPR, PRI	P
	90.5	WSMC	IIIIIIIII	CLS, NPR, PRI	P
	91.5	WAWL	III	AC	C
	102.7	WBDX	III	RLG AC	
	89.7	WDYN	IIIIIIII	RLG	C
AM	1070	WFLI	IIIIIIII	RLG	
AM	1450	WLMR	I	RLG	
AM	1490	WJOC	I	GOSP	
	94.3	WJTT	III	URBAN	
	102.3	WGOW	III	NWS, TLK	
AM	1150	WGOW	IIIII	NWS, TLK	
AM	1370	WDEF	IIIII	SPRTS	

JACKSON, TN

	FREQ	STATION	POWER	FORMAT	
	103.1	WMXX	IIIII	OLDS	
	104.1	WTNV	IIIIIIIII	CTRY	
AM	1390	WTJS	IIIII	BB, NOS	
	90.1	WKNP	III	CLS, NPR, PRI	P
	101.5	WNWS	III	NWS, TLK	
AM	1310	WDXI	IIIII	BUS NWS	

KINGSPORT, TN

	FREQ	STATION	POWER	FORMAT
	98.5	WTFM	IIIIIIIII	AC
	99.3	WAEZ	IIIIIIIII	HOT AC

65

KINGSPORT • KNOXVILLE • MEMPHIS, TN

	FREQ	STATION	POWER	FORMAT	
	92.7	WABN	III	TOP 40	
	103.9	WXIS	III	TOP 40	
	101.5	WQUT	IIIIIIII	CLS RK	
	104.9	WKOS	III	OLDS	
	96.9	WXBQ	IIIIIIII	CTRY	
	104.3	WEYE	III	CTRY	
	106.5	WJDT	III	CTRY	
AM	640	WGOC	IIIIIIII	CTRY	
AM	1050	WGAT	I	CTRY	
AM	1240	WBEJ	I	CTRY	
AM	1370	WRGS	I	CTRY	
AM	1400	WKPT	I	BB, NOS	
AM	1590	WKTP	IIIII	BB, NOS	
	89.5	WETS	IIIIIIII	DVRS, NPR, PRI	P
AM	1090	WKCV	IIIII	GOSP	
AM	1260	WMCH	I	RLG	
AM	910	WJCW	IIIIIIII	NWS, TLK	
AM	980	WXBQ	IIIIIIII	NWS, TLK	
AM	1320	WKIN	IIIII	NWS, TLK	

KNOXVILLE, TN

	FREQ	STATION	POWER	FORMAT	
	97.5	WJXB	IIIIIIII	AC	
	93.1	WWST	III	TOP 40	
	94.3	WNFZ	III	DVRS	
	100.3	WOKI	IIIIIIII	CLS RK	
	103.5	WIMZ	IIIIIIII	CLS RK	
	105.3	WESK	III	70's RK	
	102.1	WMYU	IIIII	OLDS	
	95.7	WGAP	III	CTRY	
	104.5	WQIX	III	CTRY	
	105.5	WDLY	III	CTRY	
	107.7	WIVK	IIIIIIII	CTRY	
AM	1380	WYSH	I	CTRY	
AM	1400	WGAP	I	CTRY	
	89.9	WDVX	III	CTRY	P
	90.3	WUTK	I	AC	C
	91.9	WUOT	IIIIIIII	CLS, NPR, PRI	P
	96.3	WJBZ	III	GOSP	
AM	850	WJBZ	IIIIIIII	GOSP	
AM	900	WKXV	III	GOSP	
	99.1	WNOX	III	NWS, TLK	
AM	620	WRJZ	IIIIIIII	NWS, TLK	
AM	990	WNOX	IIIIIIII	NWS, TLK	
AM	1240	WIMZ	IIIIIIII	SPRTS	
AM	1340	WKGN	I	URBAN	

MEMPHIS, TN

	FREQ	STATION	POWER	FORMAT	
	99.7	WMC	IIIII	HOT AC	
	104.5	WRVR	IIIIIIII	AC	
	107.5	WKSL	III	TOP 40	
	92.9	WMFS	III	ROCK	
	95.7	WRXQ	III	DVRS	
	98.1	WSRR	III	CLS RK	
	102.7	WEGR	IIIIIIII	CLS RK	
	107.3	KOSE	I	CLS RK	
	94.1	WOGY	IIIII	CTRY	
	96.3	KHLS	IIIII	CTRY	
	105.9	WGKX	IIIIIIII	CTRY	
	89.3	WYPL	III	NWS	P
	89.9	WEVL	III	DVRS	P
	91.1	WKNO	IIIIIIII	CLS, NPR, PRI	P
	91.7	WUMR	III	JZZ	C

66

	FREQ	STATION	POWER	FORMAT	
AM	1240	WAVN	I	GOSP	
AM	1340	WLOK	I	GOSP	
AM	1380	WOOM	III	GOSP	
AM	1480	WBBP	IIIII	GOSP	
	88.1	WURC	III	GOSP	C
AM	640	WCRV	IIIIIIII	RLG	
AM	990	KWAM	IIIIIIII	RLG	
AM	1170	WPLX	I	EZ	
	92.7	WKRA	I	URBAN	
	97.1	WHRK	IIIIIIII	URBAN	
	101.1	KJMS	IIIII	URBAN	
	107.1	KXHT	III	URBAN	
AM	560	WHBQ	IIIIIIII	SPRTS	
AM	600	WREC	IIIIIIII	NWS, TLK	
AM	680	WJCE	IIIIIIII	URBAN	
AM	790	WMC	IIIIIIII	NWS, TLK	
AM	1070	WDIA	IIIIIIII	URBAN	
AM	1430	WOWW	III	NWS, TLK	

TN

NASHVILLE, TN

	FREQ	STATION	POWER	FORMAT	
	92.9	WJXA	IIIIIIII	LGHT AC	
	107.5	WRVW	IIIIIIII	HOT AC	
	102.5	WQZQ	IIIII	TOP 40	
	94.1	WRLG	I	ALT RK	
	100.1	WRLT	III	ALT RK	
	103.3	WKDF	IIIIIIII	MOD RK	
	104.5	WGFX	III	CLS RK	
	105.9	WNRQ	IIIIIIII	CLS RK	
	96.3	WRMX	IIIIIIII	OLDS	
AM	1470	WVOL	IIIII	OLDS	
	94.3	WDBL	I	CTRY	
	95.5	WSM	IIIIIIII	CTRY	
	97.9	WSIX	IIIIIIII	CTRY	
	98.9	WANT	III	CTRY	
AM	650	WSM	IIIIIIII	CTRY	
AM	1260	WDKN	IIIII	CTRY	
AM	1160	WAMB	IIIIIIII	BB, NOS	
	88.3	WMTS	I	AC	C
	88.5	WVCP	III	DVRS	C
	89.5	WMOT	IIIIIIII	JZZ, NPR	P
	90.3	WPLN	IIIIIIII	CLS, NPR, PRI	P
	91.5	WFMQ	I	DVRS	C
	93.7	WYYB	I	DVRS	
AM	1240	WNSG	I	GOSP	
	88.1	WFSK	I	RLG	C
AM	1300	WNQM	IIIIIIII	RLG	
	92.1	WQQK	III	URBAN	
AM	880	WMDB	IIIII	URBAN	
AM	1450	WGNS	I	NWS, TLK	
AM	1510	WLAC	IIIIIIII	NWS, TLK	

88.1

FL	PENSACOLA	WUWF	57
FL	SARASOTA	WJIS	57
NC	ASHEVILLE	WCQS	61
NC	RALEIGH-DUR.	WKNC	63
SC	GREEN.-SPART.	WSBF	65
TN	CHATTANOOGA	WUTC	65
TN	MEMPHIS	WURC	67
TN	NASHVILLE	WFSK	67

88.3

TN	NASHVILLE	WMTS	67

88.5

FL	JACKSONVILLE	WFCF	55
FL	TAMPA-ST. PETE	WMNF	58
MS	HATTIESBURG	WRAS	59
MS	JACKSON	WUSM	61
NC	GREEN./WIN SAL.	WJSU	60
SC	GREEN.-SPART.	WFDD	62
TN	NASHVILLE	WEPC	65
		WVCP	67

88.7

AL	DOTHAN	WRWA	52
NC	RALEIGH-DUR.	WXDU	63

88.9

FL	FT. PIERCE	WQCS	54
FL	MIAMI-FT. LAUDERD	WDNA	56
FL	TALLAHASSEE	WFSU	57
NC	RALEIGH-DUR.	WSHA	63

89.1

FL	GAINESVILLE	WUFT	54
FL	PANAMA CITY	WFSW	57

89.3

AL	HUNTSVILLE	WLRH	52
GA	ATLANTA	WRFG	59
NC	GREEN./WIN SAL.	WSOE	62
NC	EGREN/NW BERN	WTEB	63
NC	RALEIGH-DUR.	WXYC	63
SC	CHARLESTON	WSCI	64
TN	MEMPHIS	WYPL	67

89.5

FL	MELBOURNE	WFIT	56
MS	TUPELO	WMAE	61
TN	JOHN CITY-KING	WETS	66
TN	NASHVILLE	WMOT	67

89.7

FL	TALLAHASSEE	WVFS	57
FL	TAMPA-ST. PETE	WUSF	58
TN	CHATTANOOGA	WDYN	65

89.9

FL	JACKSONVILLE	WJCT	55
FL	ORLANDO	WUCF	56
NC	CHARLOTTE	WDAV	61
TN	KNOXVILLE	WDVX	66
TN	MEMPHIS	WEVL	67

90.1

AL	HUNTSVILLE	WOCG	52
FL	FT. MYERS	WGCU	54
GA	ATLANTA	WABE	59
MS	JACKSON	WMPR	60
NC	GREEN./WIN SAL.	WNAA	62
TN	JACKSON	WKNP	66

90.3

AL	BIRMINGHAM	WBHM	52
GA	SAVANNAH	WHCJ	60
MS	GULFPORT	WMAH	60
NC	EGREN/NW BERN	WKNS	63
TN	KNOXVILLE	WUTK	66
TN	NASHVILLE	WPLN	67

90.5

FL	MIAMI-FT. LAUDERD	WVUM	56
FL	TALLAHASSEE	WAMF	58
NC	GREEN./WIN SAL.	WWCU	62
NC	GREEN./WIN SAL.	WSNC	62
TN	CHATTANOOGA	WSMC	65

90.7

AL	MONTGOMERY	WVAS	53
AL	TUSCALOOSA	WVUA	53
FL	ORLANDO	WMFE	56
FL	PANAMA CITY	WKGC	57
FL	W. PALM BEACH	WXEL	58
GA	ATLANTA	WWGC	59
GA	AUGUSTA	WACG	59
NC	CHARLOTTE	WFAE	61
NC	RALEIGH-DUR.	WNCU	63

90.9

AL	HUNTSVILLE	WJAB	52
FL	JACKSONVILLE	WKTZ	55
FL	KEY WEST	WJIR	55
NC	GREEN./WIN SAL.	WQFS	62

91.1

AL	BIRMINGHAM	WVSU	52
GA	ATLANTA	WREK	59
GA	SAVANNAH	WSVH	60
TN	MEMPHIS	WKNO	67

91.3

AL	MOBILE	WHIL	53
FL	MIAMI-FT. LAUDERD	WLRN	56
MS	JACKSON	WMPN	60
NC	EGREN/NW BERN	WZMB	63
NC	WILMINGTON	WHQR	63
SC	COLUMBIA	WLTR	64

91.5

AL	TUSCALOOSA	WUAL	53
FL	ORLANDO	WPRK	56
FL	TALLAHASSEE	WFSQ	58
NC	RALEIGH-DUR.	WUNC	63
TN	CHATTANOOGA	WAWL	65
TN	NASHVILLE	WFMQ	67

91.7

GA	ALBANY	WUNV	58
GA	COLUMBUS	WTJB	59
NC	CHARLOTTE	WSGE	61
TN	MEMPHIS	WUMR	67

91.9

GA	ATLANTA	WCLK	59
NC	FAYETTEVILLE	WFSS	62
TN	KNOXVILLE	WUOT	66

92.1

AL	DOTHAN	WJJN	52
AL	MOBILE	WZEW	53
FL	FT. WALT. BCH	WMMK	54
FL	JACKSONVILLE	WJXR	55
FL	SARASOTA	WCTQ	57
FL	W. PALM BEACH	WRLX	58
MS	HATTIESBURG	WJMG	61
SC	MYRTLE BEACH	WJYR	65
TN	NASHVILLE	WQQK	67

92.3

AL	MONTGOMERY	WLWI	53
FL	MIAMI-FT. LAUDERD	WCMQ	56
FL	ORLANDO	WWKA	56
GA	AUGUSTA	WAEG	59
GA	MACON	WPPG	60
NC	GREEN./WIN SAL.	WKRR	62
NC	EGREN/NW BERN	WQSL	62
TN	CHATTANOOGA	WDEF	65

92.5

AL	BIRMINGHAM	WZJT	52
AL	HUNTSVILLE	WWXQ	52
FL	GAINESVILLE	WNDT	54
FL	KEY WEST	WEOW	55
FL	PANAMA CITY	WPAP	57
FL	TAMPA-ST. PETE	WYUU	58
MS	TUPELO	WESE	61
SC	GREEN.-SPART.	WESC	64

92.7

FL	FT. PIERCE	WZZR	54
FL	JACKSONVILLE	WJBT	55
NC	CHARLOTTE	WCCJ	61
TN	JOHN CITY-KING	WABN	66
TN	MEMPHIS	WKRA	67

92.9

AL	MOBILE	WBLX	53
AL	TUSCALOOSA	WTUG	53
FL	GAINESVILLE	WMFQ	54
GA	ATLANTA	WZGC	59
MS	JACKSON	WJXN	61
TN	MEMPHIS	WMFS	66
TN	NASHVILLE	WJXA	67

93.1

FL	DAYTONA BCH.	WKRO	53
FL	MIAMI-FT. LAUDERD	WTMI	56
GA	ATLANTA	WEAS	60
NC	GREEN./WIN SAL.	WMQX	62
TN	KNOXVILLE	WWST	66

93.3

FL	JACKSONVILLE	WPLA	55

FL	TAMPA-ST. PETE	WFLZ	58
MS	TUPELO	WSYE	61
NC	EGREN/NW BERN	WERO	62
SC	CHARLESTON	WWWZ	64

93.5

FL	KEY WEST	WKRY	55
MS	JACKSON	WHJT	60
SC	COLUMBIA	WARQ	64

93.7

AL	BIRMINGHAM	WDJC	52
AL	DOTHAN	WRJM	52
FL	FT. PIERCE	WGYL	54
FL	GAINESVILLE	WOGK	54
MS	GULFPORT	WMJY	60
SC	GREEN.-SPART.	WFBC	64
TN	CHATTANOOGA	WMPZ	65
TN	NASHVILLE	WYYB	67

93.9

FL	MIAMI-FT. LAUDER	WLVE	56
GA	AUGUSTA	WGOR	59
MS	JACKSON	WVIV	60
NC	GREEN./WIN SAL.	WRSN	62
SC	MYRTLE BEACH	WJXY	65

94.1

AL	HUNTSVILLE	WXQW	52
FL	JACKSONVILLE	WSOS	54
FL	PENSACOLA	WMEZ	57
FL	TALLAHASSEE	WAKU	57
FL	TAMPA-ST. PETE	WSJT	58
GA	ATLANTA	WSTR	59
GA	SAVANNAH	WCHY	60
NC	GREEN./WIN SAL.	WWGL	62
NC	EGREN/NW BERN	WNBR	63
TN	MEMPHIS	WOGY	67
TN	NASHVILLE	WRLG	67

94.3

AL	DOTHAN	WIZB	52
FL	KEY WEST	WGMX	55
NC	EGREN/NW BERN	WGPM	62
SC	CHARLESTON	WSSP	64
TN	CHATTANOOGA	WJTT	65
TN	KNOXVILLE	WNFZ	66
TN	NASHVILLE	WDBL	67

94.5

AL	BIRMINGHAM	WYSF	52
MS	GULFPORT	WJZD	60
SC	GREEN.-SPART.	WMUU	65
SC	MYRTLE BEACH	WRNN	65

94.7

FL	FT. PIERCE	WBBE	54
MS	JACKSON	WTYX	60
NC	RALEIGH-DUR.	WQDR	63

94.9

AL	MOBILE	WKSJ	53
FL	MIAMI-FT. LAUDER	WZTA	56
FL	TALLAHASSEE	WTNT	57
FL	TAMPA-ST. PETE	WWRM	58
GA	ATLANTA	WPCH	59
SC	MYRTLE BEACH	WVCO	65

95.1

AL	HUNTSVILLE	WNDA	52
AL	MONTGOMERY	WXFX	53
FL	JACKSONVILLE	WAPE	54
FL	MELBOURNE	WBVD	55
GA	AUGUSTA	WCHZ	59
NC	CHARLOTTE	WNKS	61
NC	EGREN/NW BERN	WRNS	63
SC	CHARLESTON	WSSX	64

95.3

AL	BIRMINGHAM	WFFN	52
FL	FT. MYERS	WOLZ	54
FL	ORLANDO	WTLN	57
MS	TUPELO	WAFM	61
SC	COLUMBIA	WFMV	64
AL	DOTHAN	WTVY	52
FL	GAINESVILLE	WNDD	54
GA	SAVANNAH	WIXV	60
MS	JACKSON	WKTF	60
TN	NASHVILLE	WSM	67

95.5

AL	TUSCALOOSA	WBHJ	53
FL	DAYTONA BCH.	WHOG	53
FL	MIAMI-FT. LAUDERD	WXDJ	56

FL	TAMPA-ST. PETE	WSSR	58
NC	FAYETTEVILLE	WKML	62
TN	KNOXVILLE	WGAP	66
TN	MEMPHIS	WRXQ	66

95.9

| FL | PANAMA CITY | WRBA | 57 |

96.1

AL	MOBILE	WRKH	53
AL	MONTGOMERY	WRWO	53
FL	FT. MYERS	WRXK	53
FL	JACKSONVILLE	WEJZ	54
FL	TALLAHASSEE	WHBX	58
FL	TAMPA-ST. PETE	WGUL	58
GA	ATLANTA	WKLS	59
NC	RALEIGH-DUR.	WBBB	63
SC	CHARLESTON	WAVF	64

96.3

GA	ALBANY	WJIZ	59
GA	AUGUSTA	WRXR	59
MS	JACKSON	WJDX	60
NC	EGREN/NW BERN	WRHT	62
TN	KNOXVILLE	WJBZ	66
TN	MEMPHIS	KHLS	66
TN	NASHVILLE	WRMX	67

96.5

AL	BIRMINGHAM	WMJJ	52
FL	MIAMI-FT. LAUD.	WPOW	56
FL	ORLANDO	WHTQ	56
GA	MACON	WIBD	60
GA	SAVANNAH	WJCL	60
NC	ASHEVILLE	WZLS	61
TN	CHATTANOOGA	WDOD	65

96.7

MS	GULFPORT	WLRK	60
MS	TUPELO	WSEL	61
SC	COLUMBIA	WLTY	64

96.9

AL	DOTHAN	WDJR	52
AL	HUNTSVILLE	WRSA	52
FL	FT. MYERS	WINK	53
FL	JACKSONVILLE	WKQL	55
SC	CHARLESTON	WSUY	64
TN	JOHN CITY-KING	WXBQ	66

97.1

AL	MONTGOMERY	WMCZ	53
FL	FT. PIERCE	WOSN	54
FL	TAMPA-ST. PETE	WLVU	58
NC	GREEN./WIN SAL.	WQMG	62
TN	MEMPHIS	WHRK	67

97.3

FL	GAINESVILLE	WSKY	54
FL	MIAMI-FT. LAUD.	WFLC	56
GA	SAVANNAH	WAEV	60
NC	WILMINGTON	WMNX	63
TN	CHATTANOOGA	WKXJ	65
AL	MOBILE	WABB	53

97.5

FL	LAKELAND	WPCV	55
GA	ATLANTA	WHTA	59
SC	COLUMBIA	WCOS	64
TN	KNOXVILLE	WJXB	66

97.7

AL	BIRMINGHAM	WKLD	52
MS	JACKSON	WRJH	61
NC	EGREN/NW BERN	WZBR	63
SC	MYRTLE BEACH	WWXM	65

97.9

FL	JACKSONVILLE	WFSJ	55
FL	TAMPA-ST. PETE	WXTB	58
FL	W. PALM BEACH	WRMF	58
GA	MACON	WIBB	60
NC	CHARLOTTE	WPEG	62
TN	NASHVILLE	WSIX	67

98.1

MS	HATTIESBURG	WMXI	61
NC	FAYETTEVILLE	WQSM	62
SC	CHARLESTON	WYBB	64
TN	CHATTANOOGA	WXKT	65
TN	MEMPHIS	WSRR	67

98.3

AL	MOBILE	WDLT	53
FL	MIAMI-FT. LAUD.	WRTO	56
GA	AUGUSTA	WSLT	59

GA	COLUMBUS	WAGH	59
NC	GREEN./WIN SAL.	WIST	62
NC	EGREN/NW BERN	WCZI	63

98.5
FL	FT. MYERS	WDRR	54
FL	PANAMA CITY	WFSY	57
GA	ATLANTA	WSB	59
MS	TUPELO	WZLQ	61
SC	COLUMBIA	WLXC	64
SC	FLORENCE	WHSC	64
SC	MYRTLE BEACH	WDAI	65
TN	JOHN CITY-KING	WTFM	66

98.7
AL	BIRMINGHAM	WBHK	52
FL	KEY WEST	WCNK	55
FL	SARASOTA	WLLD	57
NC	GREEN./WIN SAL.	WKSI	62
NC	EGREN/NW BERN	WKOO	62

98.9
AL	MONTGOMERY	WBAM	53
FL	ORLANDO	WMMO	56
FL	TALLAHASSEE	WBZE	57
SC	CHARLESTON	WWBZ	64
SC	GREEN.-SPART.	WSPA	64
TN	NASHVILLE	WANT	67

99.1
AL	HUNTSVILLE	WAHR	52
FL	JACKSONVILLE	WQIK	55
FL	MIAMI-FT. LAUD.	WEDR	56
GA	MACON	WAYS	60
MS	GULFPORT	WKNN	60
TN	KNOXVILLE	WNOX	66

99.3
FL	FT. MYERS	WJBX	53
FL	MELBOURNE	WLRQ	55
FL	PANAMA CITY	WSHF	57
MS	TUPELO	WBIP	61
TN	JOHN CITY-KING	WAEZ	66

99.5
AL	BIRMINGHAM	WZRR	52
FL	FT. WALT. BCH	WKSM	54
FL	KEY WEST	WAIL	55
FL	TAMPA-ST. PETE	WQYK	58
FL	W. PALM BEACH	WJBW	58
GA	AUGUSTA	WKXC	59
NC	GREEN./WIN SAL.	WMAG	62
NC	EGREN/NW BERN	WXNR	62
SC	MYRTLE BEACH	WMYB	65

99.7
AL	DOTHAN	WOOF	52
FL	FT. PIERCE	WPAW	54
GA	ATLANTA	WNNX	59
MS	JACKSON	WJMI	61
NC	CHARLOTTE	WRFX	61
TN	MEMPHIS	WMC	66

99.9
AL	MOBILE	WMXC	53
FL	TALLAHASSEE	WWFO	57
NC	ASHEVILLE	WKSF	61

100.1
FL	PANAMA CITY	WPCF	57
GA	COLUMBUS	WGSY	59
SC	COLUMBIA	WSCQ	64
SC	FLORENCE	WWFN	64
TN	NASHVILLE	WRLT	67

100.3
FL	KEY WEST	WCTH	55
FL	ORLANDO	WSHE	56
MS	HATTIESBURG	WNSL	61
NC	GREEN./WIN SAL.	WHSL	62
TN	KNOXVILLE	WOKI	66

100.5
| AL | DOTHAN | WXUS | 52 |
| SC | CHARLESTON | WLLC | 64 |

100.7
AL	TUSCALOOSA	WLXY	53
FL	MIAMI-FT. LAUD.	WHYI	56
FL	PENSACOLA	WWRO	57
FL	TAMPA-ST. PETE	WAKS	58
SC	MYRTLE BEACH	WGTN	65

100.9
| FL | GAINESVILLE | WYGC | 54 |
| GA | MACON | WPGA | 60 |

101.1
AL	DOTHAN	WZTZ	52
FL	PANAMA CITY	WYOO	57
NC	GREEN./WIN SAL.	WKXU	62
NC	EGREN/NW BERN	WKJA	62
SC	GREEN.-SPART.	WROQ	64
TN	MEMPHIS	KJMS	67

101.3
| AL | DOTHAN | WAGF | 52 |
| NC | WILMINGTON | WWQQ | 63 |

101.5
FL	MIAMI-FT. LAUD.	WLYF	56
FL	PENSACOLA	WTKX	57
FL	TALLAHASSEE	WXSR	57
GA	ATLANTA	WKHX	59
NC	RALEIGH-DUR.	WRAL	63
TN	JACKSON	WNWS	66
TN	JOHN CITY-KING	WQUT	66

101.7
FL	FT. PIERCE	WAVW	54
GA	ALBANY	WKAK	58
GA	AUGUSTA	WTHO	59
GA	MACON	WRBV	60
MS	JACKSON	WYOY	60
SC	CHARLESTON	WMGL	64
SC	MYRTLE BEACH	WKZQ	65

101.9
AL	MONTGOMERY	WJCC	53
FL	FT. MYERS	WWGR	54
MS	TUPELO	WFTA	61
NC	CHARLOTTE	WBAV	61
NC	EGREN/NW BERN	WIKS	63
TN	CHATTANOOGA	WSGC	65

102.1
AL	HUNTSVILLE	WDRM	52
AL	MOBILE	WHXT	53
FL	KEY WEST	WKLG	55
GA	SAVANNAH	WZAT	60
SC	FLORENCE	WMXT	64
TN	KNOXVILLE	WMYU	66

102.3
FL	FT. PIERCE	WMBX	54
FL	GAINESVILLE	WTRS	54
GA	AUGUSTA	WEKL	59
MS	GULFPORT	WGCM	60
SC	COLUMBIA	WMFX	64
TN	CHATTANOOGA	WGOW	65

102.5
AL	BIRMINGHAM	WOWC	52
AL	DOTHAN	WESP	52
FL	KEY WEST	WPIK	55
GA	MACON	WLCG	60
MS	HATTIESBURG	WJKX	61
NC	RALEIGH-DUR.	WHLQ	63
SC	CHARLESTON	WXLY	64
TN	NASHVILLE	WQZQ	67

102.7
FL	MELBOURNE	WHKR	55
FL	MIAMI-FT. LAUD.	WMXJ	56
FL	PENSACOLA	WXBM	57
GA	AUGUSTA	WAJY	59
NC	WILMINGTON	WGNI	63
TN	CHATTANOOGA	WBDX	65
TN	MEMPHIS	WEGR	67

102.9
FL	JACKSONVILLE	WMXQ	54
GA	COLUMBUS	WVRK	59
MS	JACKSON	WMSI	60
NC	EGREN/NW BERN	WELS	63
SC	FLORENCE	WSQN	64

103.1
FL	KEY WEST	WFKZ	55
FL	ORLANDO	WLOQ	56
FL	TALLAHASSEE	WAIB	57
GA	AUGUSTA	WFXA	59
MS	GULFPORT	WOSM	60
SC	COLUMBIA	WOMG	64
SC	GREEN.-SPART.	WRIX	64
SC	MYRTLE BEACH	WYAK	65
TN	JACKSON	WMXX	66

103.3
| AL | MONTGOMERY | WMXS | 53 |

C

FL	DAYTONA BCH.	WVYB	53			
GA	ATLANTA	WVEE	59			
NC	EGREN/NW BERN	WMGV	62			
SC	GREEN.-SPART.	WOLT	64			
TN	NASHVILLE	WKDF	67			

103.5

FL	MIAMI-FT. LAUD.	WPLL	56
FL	PANAMA CITY	WMXP	57
FL	SARASOTA	WDUV	57
GA	ALBANY	WJAD	58
MS	TUPELO	WRBO	61
SC	CHARLESTON	WEZL	64
TN	KNOXVILLE	WIMZ	66

103.7

FL	FT. PIERCE	WQOL	54
FL	GAINESVILLE	WRUF	54
MS	HATTIESBURG	WHER	61
NC	CHARLOTTE	WSOC	61
NC	WILMINGTON	WLTT	63

103.9

AL	DOTHAN	WQLS	52
FL	FT. MYERS	WXKB	53
GA	SAVANNAH	WSGF	60
NC	RALEIGH-DUR.	WNNL	63
SC	GREEN.-SPART.	WOLI	64
TN	JOHN CITY-KING	WXIS	66

104.1

FL	TALLAHASSEE	WGLF	57
NC	GREEN./WIN SAL.	WTQR	62
SC	MYRTLE BEACH	WYAV	65
TN	JACKSON	WTNV	66

104.3

AL	HUNTSVILLE	WZYP	52
FL	W. PALM BEACH	WEAT	58
GA	AUGUSTA	WBBQ	59
TN	JOHN CITY-KING	WEYE	66

104.5

FL	JACKSONVILLE	WFYV	55
GA	ALBANY	WGPC	58
MS	HATTIESBURG	WXRR	61
NC	WILMINGTON	WRQR	63
SC	CHARLESTON	WRFQ	64
TN	KNOXVILLE	WQIX	66
TN	MEMPHIS	WRVR	66
TN	NASHVILLE	WGFX	67

104.7

AL	BIRMINGHAM	WZZK	52
FL	FT. PIERCE	WFLM	54
FL	FT. WALT. BCH	WAAZ	54
FL	KEY WEST	WWUS	55
FL	TAMPA-ST. PETE	WRBQ	58
NC	CHARLOTTE	WSSS	61
SC	COLUMBIA	WNOK	64

104.9

FL	GAINESVILLE	WRKG	54
FL	TALLAHASSEE	WFLV	57
GA	COLUMBUS	WFXE	60
MS	GULFPORT	WYOK	60
SC	GREEN.-SPART.	WCCP	65
TN	JOHN CITY-KING	WKOS	66

105.1

FL	MIAMI-FT. LAUD.	WHQT	56
FL	ORLANDO	WOMX	56
FL	PANAMA CITY	WAKT	57
NC	EGREN/NW BERN	WANG	63
NC	RALEIGH-DUR.	WDCG	63

105.3

FL	GAINESVILLE	WYKS	54
GA	MACON	WDEN	60
GA	SAVANNAH	WRHQ	60
MS	TUPELO	WWKZ	61
SC	CHARLESTON	WNST	64
TN	KNOXVILLE	WESK	66

105.5

AL	MOBILE	WNSP	53
AL	TUSCALOOSA	WRTR	53
FL	FT. MYERS	WQNU	54
FL	FT. WALT. BCH	WYZB	54
FL	JACKSONVILLE	WJQR	55
FL	KEY WEST	WAVK	55
FL	TAMPA-ST. PETE	WTBT	58
FL	W. PALM BEACH	WWLV	58
NC	EGREN/NW BERN	WXQR	62

SC	FLORENCE	WDAR	64
TN	CHATTANOOGA	WLMX	65
TN	KNOXVILLE	WDLY	66

105.7

FL	JACKSONVILLE	WXQL	55
GA	ATLANTA	WGST	59
GA	AUGUSTA	WZNY	59

105.9

AL	BIRMINGHAM	WENN	52
FL	DAYTONA BCH.	WOCL	53
FL	MIAMI-FT. LAUD.	WBGG	56
FL	PANAMA CITY	WILN	57
MS	GULFPORT	WXRG	60
SC	MYRTLE BEACH	WNMB	65
TN	MEMPHIS	WGKX	67
TN	NASHVILLE	WNRQ	67

106.1

AL	HUNTSVILLE	WTAK	52
FL	TALLAHASSEE	WWLD	57
NC	CHARLOTTE	WNMX	61
SC	CHARLESTON	WTUA	64

106.3

FL	FT. MYERS	WJST	54
FL	KEY WEST	WZMQ	55
GA	MACON	WQBZ	60
MS	HATTIESBURG	WMFM	61
NC	WILMINGTON	WCCA	63
SC	FLORENCE	WYNN	64

106.5

AL	MOBILE	WAVH	53
FL	JACKSONVILLE	WBGB	55
FL	SARASOTA	WSRZ	57
GA	ALBANY	WZIQ	59
NC	CHARLOTTE	WEND	61
NC	EGREN/NW BERN	WSFL	62
SC	MYRTLE BEACH	WSYN	65
TN	CHATTANOOGA	WSKZ	65
TN	JOHN CITY-KING	WJDT	66

106.7

AL	DOTHAN	WKMX	52
FL	MIAMI-FT. LAUD.	WRMA	56
MS	TUPELO	WWZD	61

106.9

AL	BIRMINGHAM	WODL	52

107.1

FL	FT. MYERS	WCKT	54
FL	KEY WEST	WIIS	55
FL	MELBOURNE	WAOA	55
MS	GULFPORT	WXYK	60
NC	RALEIGH-DUR.	WFXC	63
TN	MEMPHIS	KXHT	67

107.3

FL	JACKSONVILLE	WROO	55
FL	PENSACOLA	WYCL	57
FL	TAMPA-ST. PETE	WCOF	58
GA	COLUMBUS	WCGQ	59
NC	EGREN/NW BERN	WTKF	63
SC	GREEN.-SPART.	WJMZ	65
TN	MEMPHIS	KOSE	67

107.5

FL	MIAMI-FT. LAUD.	WAMR	56
NC	GREEN./WIN SAL.	WKZL	62
NC	WILMINGTON	WSFM	63
SC	CHARLESTON	WNKT	64
TN	MEMPHIS	WKSL	66
TN	NASHVILLE	WRVW	67

107.7

AL	BIRMINGHAM	WRAX	52
GA	AUGUSTA	WUUS	59
NC	FAYETTEVILLE	WUKS	62
TN	KNOXVILLE	WIVK	66

107.9

FL	KEY WEST	WVMQ	55
FL	PANAMA CITY	WDRK	57
FL	SARASOTA	WYNF	57
FL	W. PALM BEACH	WIRK	58
GA	MACON	WPEZ	60
NC	CHARLOTTE	WLNK	61
NC	EGREN/NW BERN	WNCT	63
SC	MYRTLE BEACH	WGTR	65
TN	CHATTANOOGA	WOGT	65

C

AM

540

FL	ORLANDO	WQTM	57
GA	COLUMBUS	WDAK	60
NC	RALEIGH-DUR.	WETC	63
SC	FLORENCE	WYNN	64

550

FL	JACKSONVILLE	WAYR	55

560

AL	DOTHAN	WOOF	52
FL	MIAMI-FT. LAUD.	WQAM	56
SC	COLUMBIA	WVOC	64
TN	MEMPHIS	WHBQ	67

570

FL	TAMPA-ST. PETE	WHNZ	58
MS	GULFPORT	WVMI	60
NC	ASHEVILLE	WWNC	61

580

FL	ORLANDO	WDBO	57
GA	AUGUSTA	WGAC	59
MS	TUPELO	WELO	61

590

FL	PANAMA CITY	WDIZ	57

600

FL	JACKSONVILLE	WBWL	55
NC	GREEN./WIN SAL.	WSJS	62
TN	MEMPHIS	WREC	67

610

AL	BIRMINGHAM	WEZN	52
FL	MIAMI-FT. LAUD.	WIOD	56
FL	PENSACOLA	WVTJ	57
NC	CHARLOTTE	WFNZ	62

620

FL	TAMPA-ST. PETE	WSUN	58
MS	JACKSON	WJDS	61
NC	RALEIGH-DUR.	WDNC	63
SC	COLUMBIA	WTGH	64
TN	KNOXVILLE	WRJZ	66

630

GA	SAVANNAH	WBMQ	60
NC	WILMINGTON	WMFD	63

640

FL	W. PALM BEACH	WLVJ	58
GA	ATLANTA	WGST	59
NC	FAYETTEVILLE	WFNC	62
TN	JOHN CITY-KING	WGOC	66
TN	MEMPHIS	WCRV	67

650

TN	NASHVILLE	WSM	67

660

AL	MOBILE	WDLT	53

670

FL	MIAMI-FT. LAUD.	WWFE	56

680

GA	ATLANTA	WCNN	59
NC	RALEIGH-DUR.	WPTF	63
TN	MEMPHIS	WJCE	67

690

FL	JACKSONVILLE	WOKV	55

710

AL	MOBILE	WNTM	53
FL	MIAMI-FT. LAUD.	WAQI	56

730

AL	HUNTSVILLE	WUMP	52
SC	CHARLESTON	WPAL	64

740

AL	MONTGOMERY	WMSP	53
FL	ORLANDO	WWNZ	57
FL	W. PALM BEACH	WSBR	58

750

GA	ATLANTA	WSB	59

760

FL	TAMPA-ST. PETE	WBDN	58

770

AL	HUNTSVILLE	WVNN	52
FL	FT. MYERS	WWCN	54

790

FL	MIAMI-FT. LAUD.	WAXY	56
GA	ATLANTA	WQXI	59
TN	MEMPHIS	WMC	67

800

AL	MONTGOMERY	WMGY	53

820

FL	TAMPA-ST. PETE	WZTM	58

830

FL	MIAMI-FT. LAUD.	WACC	56

850

AL	BIRMINGHAM	WYDE	52
FL	GAINESVILLE	WRUF	54
FL	W. PALM BEACH	WDJA	58
NC	RALEIGH-DUR.	WRBZ	63
TN	KNOXVILLE	WJBZ	66

860

FL	MELBOURNE	WRFB	56
FL	TAMPA-ST. PETE	WGUL	58

880

TN	NASHVILLE	WMDB	67

900

AL	BIRMINGHAM	WATV	52
AL	MOBILE	WGOK	53
GA	SAVANNAH	WEAS	60
TN	KNOXVILLE	WKXV	66

910

FL	TAMPA-ST. PETE	WFNS	58
SC	CHARLESTON	WTMZ	64
SC	GREEN.-SPART.	WORD	65
TN	JOHN CITY-KING	WJCW	66

920

FL	MELBOURNE	WMEL	56

930

FL	JACKSONVILLE	WNZS	55
FL	SARASOTA	WKXY	57
NC	GREEN/NW BERN	WDLX	63

940

FL	MIAMI-FT. LAUD.	WINZ	56
GA	MACON	WMAC	60
MS	TUPELO	WCPC	61

950

MS	HATTIESBURG	WBKH	61
SC	GREEN.-SPART.	WSPA	65

960

AL	BIRMINGHAM	WERC	52
GA	ALBANY	WJYZ	58

970

FL	TAMPA-ST. PETE	WFLA	58
GA	ATLANTA	WNIV	59
SC	FLORENCE	WJMX	64

980

FL	GAINESVILLE	WLUS	54
FL	MIAMI-FT. LAUD.	WHSR	56
FL	PENSACOLA	WRNE	57
GA	MACON	WPGA	60
NC	GREEN./WIN SAL.	WAAA	62
NC	WILMINGTON	WAAV	63
TN	CHATTANOOGA	WLMX	65
TN	JOHN CITY-KING	WXBQ	66

990

FL	ORLANDO	WHOO	56
TN	KNOXVILLE	WNOX	66
TN	MEMPHIS	KWAM	67

1000

AL	HUNTSVILLE	WDJL	52

1010

FL	TAMPA-ST. PETE	WQYK	58
NC	EGREN/NW BERN	WELS	63

1020

SC	GREEN.-SPART.	WRIX	65

1030

FL	ORLANDO	WONQ	56

1040

FL	W. PALM BEACH	WJNO	58

1050

NC	CHARLOTTE	WLON	61
SC	MYRTLE BEACH	WJXY	65
TN	JOHN CITY-KING	WGAT	66

1060

FL	MELBOURNE	WAMT	56

1070

AL	BIRMINGHAM	WAPI	52
NC	GREEN./WIN SAL.	WGOS	62
NC	EGREN/NW BERN	WNCT	63
SC	GREEN.-SPART.	WHYZ	65
TN	CHATTANOOGA	WFLI	65
TN	MEMPHIS	WDIA	67

1080			
FL	MIAMI-FT. LAUD.	WVCG	56
FL	ORLANDO	WFIV	56
1090			
NC	RALEIGH-DUR.	WBZB	63
TN	JOHN CITY-KING	WKCV	66
1110			
NC	CHARLOTTE	WBT	62
1130			
FL	LAKELAND	WWBF	55
1140			
FL	MIAMI-FT. LAUD.	WQBA	56
1150			
FL	DAYTONA BCH.	WNDB	53
FL	TAMPA-ST. PETE	WTMP	58
TN	CHATTANOOGA	WGOW	65
1160			
NC	FAYETTEVILLE	WYRU	62
TN	NASHVILLE	WAMB	67
1170			
AL	MONTGOMERY	WACV	53
FL	MIAMI-FT. LAUD.	WAVS	56
TN	MEMPHIS	WPLX	67
1180			
MS	JACKSON	WJNT	61
1190			
TN	CHATTANOOGA	WSDQ	65
1200			
FL	FT. MYERS	WTLQ	54
1210			
FL	MIAMI-FT. LAUD.	WNMA	56
1220			
FL	SARASOTA	WQSA	57
1230			
AL	HUNTSVILLE	WBHP	52
AL	TUSCALOOSA	WTBC	53
FL	DAYTONA BCH.	WSBB	53
FL	GAINESVILLE	WGGG	54
FL	LAKELAND	WONN	55
FL	W. PALM BEACH	WJNA	58
GA	AUGUSTA	WKIM	59
GA	SAVANNAH	WSOK	60
NC	ASHEVILLE	WSKY	61
NC	FAYETTEVILLE	WFAI	62
NC	GREEN./WIN SAL.	WMFR	62
SC	COLUMBIA	WOIC	64
SC	FLORENCE	WOLS	64
1240			
FL	FT. MYERS	WINK	54
FL	JACKSONVILLE	WFOY	55
FL	MELBOURNE	WMMB	55
GA	MACON	WDDO	60
NC	CHARLOTTE	WHVN	62
NC	EGREN/NW BERN	WJNC	63
NC	RALEIGH-DUR.	WPJL	63
SC	MYRTLE BEACH	WLSC	65
TN	JOHN CITY-KING	WBEJ	66
TN	KNOXVILLE	WIMZ	66
TN	MEMPHIS	WAVN	67
TN	NASHVILLE	WNSG	67
1250			
FL	TAMPA-ST. PETE	WDAE	58
NC	EGREN/NW BERN	WGHB	63
SC	CHARLESTON	WTMA	64
1260			
FL	FT. WALT. BCH	WFTW	54
FL	MIAMI-FT. LAUD.	WSUA	56
NC	GREEN./WIN SAL.	WKXR	62
TN	JOHN CITY-KING	WMCH	66
TN	NASHVILLE	WDKN	67
1270			
AL	MOBILE	WKSJ	53
FL	ORLANDO	WRLZ	56
FL	TALLAHASSEE	WNLS	58
NC	CHARLOTTE	WCGC	62
NC	RALEIGH-DUR.	WMPM	63
1280			
AL	TUSCALOOSA	WWPG	53
FL	JACKSONVILLE	WSVE	55
1290			
FL	GAINESVILLE	WTMC	54
FL	W. PALM BEACH	WBZT	58
NC	EGREN/NW BERN	WJCV	63

1300			
FL	KEY WEST	WFFG	55
FL	MELBOURNE	WXXU	56
MS	JACKSON	WOAD	61
SC	GREEN.-SPART.	WCKI	65
TN	NASHVILLE	WNQM	67
1310			
AL	MOBILE	WHEP	53
NC	ASHEVILLE	WISE	61
TN	CHATTANOOGA	WDOD	65
TN	JACKSON	WDXI	66
1320			
AL	BIRMINGHAM	WAGG	52
FL	JACKSONVILLE	WJGR	55
FL	MIAMI-FT. LAUD.	WLQY	56
FL	SARASOTA	WAMR	57
NC	GREEN./WIN SAL.	WTCK	62
SC	COLUMBIA	WISW	64
TN	JOHN CITY-KING	WKIN	66
1330			
FL	FT. PIERCE	WJNX	54
FL	LAKELAND	WWAB	55
FL	PENSACOLA	WEBY	57
MS	TUPELO	WFTO	61
SC	GREEN.-SPART.	WYRD	65
SC	MYRTLE BEACH	WPJS	65
1340			
FL	DAYTONA BCH.	WROD	53
FL	FT. WALT. BCH	WFSH	54
FL	W. PALM BEACH	WPBR	58
GA	ATLANTA	WALR	59
GA	AUGUSTA	WBBQ	59
GA	COLUMBUS	WOKS	59
MS	HATTIESBURG	WAML	61
NC	CHARLOTTE	WRHI	62
NC	GREEN./WIN SAL.	WPOL	62
NC	EGREN/NW BERN	WOOW	63
SC	CHARLESTON	WQSC	64
TN	KNOXVILLE	WKGN	66
TN	MEMPHIS	WLOK	67
1350			
FL	MELBOURNE	WMMV	55
GA	MACON	WCOP	60
1360			
AL	MOBILE	WMOB	53
FL	LAKELAND	WHNR	55
FL	MIAMI-FT. LAUD.	WKAT	56
NC	RALEIGH-DUR.	WCHL	63
1370			
FL	FT. PIERCE	WAXE	54
FL	PENSACOLA	WCOA	57
NC	CHARLOTTE	WLTC	62
TN	CHATTANOOGA	WDEF	65
TN	JOHN CITY-KING	WRGS	66
1380			
FL	DAYTONA BCH.	WELE	53
FL	TAMPA-ST. PETE	WRBQ	58
FL	W. PALM BEACH	WLVS	58
GA	ATLANTA	WAOK	59
TN	KNOXVILLE	WYSH	66
TN	MEMPHIS	WOOM	67
1390			
MS	GULFPORT	WROA	60
SC	CHARLESTON	WXTC	64
TN	JACKSON	WTJS	66
1400			
AL	BIRMINGHAM	WJLD	52
FL	FT. PIERCE	WIRA	54
FL	FT. WALT. BCH	WFAV	54
FL	JACKSONVILLE	WZAZ	55
FL	MIAMI-FT. LAUD.	WFTL	56
MS	HATTIESBURG	WFOR	61
MS	JACKSON	WKXI	61
NC	GREEN./WIN SAL.	WKEW	62
SC	COLUMBIA	WCOS	64
SC	GREEN.-SPART.	WKDY	64
TN	JOHN CITY-KING	WKPT	66
TN	KNOXVILLE	WGAP	66
1410			
AL	MOBILE	WLVV	53
FL	FT. MYERS	WMYR	54
NC	RALEIGH-DUR.	WSRC	63
1420			
AL	TUSCALOOSA	WACT	53

C

FL	JACKSONVILLE	WAOC	55
FL	SARASOTA	WBRD	57
FL	W. PALM BEACH	WDBF	58
GA	COLUMBUS	WRCG	60
1430			
FL	GAINESVILLE	WWLO	54
FL	LAKELAND	WLKF	55
FL	PANAMA CITY	WLTG	57
TN	MEMPHIS	WOWW	67
1440			
AL	MONTGOMERY	WHHY	53
FL	FT. MYERS	WWCL	54
FL	ORLANDO	WPRD	56
NC	GREEN./WIN SAL.	WLXN	62
SC	GREEN.-SPART.	WGVL	65
1450			
AL	DOTHAN	WWNT	52
AL	HUNTSVILLE	WTKI	52
FL	FT. PIERCE	WSTU	54
FL	PENSACOLA	WBSR	57
FL	TALLAHASSEE	WTAL	58
GA	ALBANY	WGPC	59
NC	EGREN/NW BERN	WNOS	63
SC	CHARLESTON	WQNT	64
SC	FLORENCE	WHSC	64
TN	CHATTANOOGA	WLMR	65
TN	NASHVILLE	WGNS	67
1460			
FL	JACKSONVILLE	WZNZ	55
FL	LAKELAND	WBAR	55
GA	COLUMBUS	WPNX	59
NC	ASHEVILLE	WHBK	61
1470			
FL	TAMPA-ST. PETE	WRTB	58
MS	TUPELO	WNAU	61
SC	COLUMBIA	WQXL	64
SC	MYRTLE BEACH	WLMC	65
TN	NASHVILLE	WVOL	67
1480			
AL	MOBILE	WABB	52
FL	PANAMA CITY	WKGC	57
GA	ATLANTA	WYZE	59
GA	AUGUSTA	WRDW	59
NC	CHARLOTTE	WGFY	62
TN	MEMPHIS	WBBP	67
1490			
FL	DAYTONA BCH.	WXVQ	53
FL	FT. PIERCE	WTTB	54
FL	MIAMI-FT. LAUD.	WMBM	56
FL	SARASOTA	WWPR	57
MS	GULFPORT	WXBD	60
NC	CHARLOTTE	WSTP	62
NC	FAYETTEVILLE	WAZZ	62
NC	RALEIGH-DUR.	WDUR	63
NC	WILMINGTON	WWIL	63
TN	CHATTANOOGA	WJOC	65
1500			
FL	KEY WEST	WKIZ	55
1510			
FL	MELBOURNE	WWBC	56
TN	NASHVILLE	WLAC	67
1520			
FL	MIAMI-FT. LAUD.	WEXY	56
1530			
FL	SARASOTA	WENG	57
1550			
AL	HUNTSVILLE	WLOR	52
FL	TAMPA-ST. PETE	WAMA	58
GA	ATLANTA	WAZX	59
GA	AUGUSTA	WTHB	59
1560			
FL	MELBOURNE	WTMS	55
1580			
GA	COLUMBUS	WEAM	59
SC	GREEN.-SPART.	WDAB	65
1590			
FL	TAMPA-ST. PETE	WRXB	58
GA	ALBANY	WALG	59
MS	JACKSON	WZRX	61
TN	JOHN CITY-KING	WKTP	66
1600			
AL	HUNTSVILLE	WEUP	52
AL	MONTGOMERY	WXVI	53

FL	KEY WEST	WKWF	55
FL	ORLANDO	WOKB	57
FL	W. PALM BEACH	WPOM	58
NC	CHARLOTTE	WGIV	61
NC	FAYETTEVILLE	WIDU	62

C

MINNESOTA

DULUTH	80
MINN.-ST. PAUL	81
ROCHESTER	81
ST. CLOUD	82

NORTH DAKOTA

BISMARCK	83
FARGO-MOORHEAD	84
GRAND FORKS	84

IOWA

CEDAR FALLS	77
CEDAR RAPIDS	77
DES MOINES	77
DUBUQUE	77
DAVENPORT	78
SIOUX CITY	78

SOUTH DAKOTA

RAPID CITY	86
SIOUX FALLS	86

NEBRASKA

LINCOLN	84
OMAHA	85

KANSAS

TOPEKA	78
WICHITA	78

OKLAHOMA

LAWTON	85
OKLAHOMA CITY	85
TULSA	85

NORTH DAKOTA

MINNESOTA

SOUTH DAKOTA

IOWA

NEBRASKA

MISSOURI

KANSAS

OKLAHOMA

ARKANSAS

0   100   200 Miles

TEXAS

LOUISIANA

TEXAS

ABILENE	86
AMARILLO	86
AUSTIN	86
BEAUMONT	87
BRYAN-COLL. STAT.	87
CORPUS CHRISTI	88
DALLAS-FT. WORTH	88
EL PASO	89
HOUSTON-GALV.	89
KILLEEN-TEMPLE	90
LAREDO	90
LONGVIEW	90
LUBBOCK	90
BROWNSVILLE	91
ODESSA-MIDLAND	91
SAN ANGELO	91
SAN ANTONIO	91
TEXARKANA	92
WACO	92
WICHITA FALLS	92

MISSOURI

COLUMBIA	82
JOPLIN	82
KANSAS CITY	82
SPRINGFIELD	83
ST. LOUIS	83

ARKANSAS

FAYETTEVILLE	76
FT. SMITH	76
LITTLE ROCK	76

LOUISIANA

ALEXANDRIA	78
BATON ROUGE	79
LAFAYETTE	79
LAKE CHARLES	79
MONRO	80
NEW ORLEAN	80
SHREVEPORT	80

FREQ	STATION	POWER	FORMAT	
101.9	KMXF	IIIIIIII	HOT AC	
107.9	KEZA	IIIIIIII	AC	
105.7	KMCK	IIIII	TOP 40	
92.1	KKEG	IIIII	ROCK	
104.9	KBRS	III	DVRS	
94.3	KAMO	IIIIIIII	OLDS	
98.3	KFAY	IIIII	CTRY	
103.9	KKIX	IIIII	CTRY	
AM 1290	KUOA	IIIII	CTRY	
106.5	KBVA	IIIIIIII	BB, NOS	
91.3	KUAF	IIIIIIII	CLS, NPR, PRI	P
94.9	KDAB	IIIIIIII	GOSP	
101.1	KLRC	III	RLG AC	C
AM 790	KURM	IIIIIIII	NWS, TLK	
AM 1030	KFAY	IIIII	NWS, TLK	

FT. SMITH, AR

95.9	KMXJ	IIIIIIII	HOT AC	
93.7	KISR	IIIIIIII	TOP 40	
97.9	KZBB	IIIIIIII	TOP 40	
92.3	KREU	III	CLS RK	
102.7	KLSZ	III	CLS RK	
100.7	KBBQ	III	OLDS	
92.5	KPRV	III	CTRY	
99.1	KMAG	IIIIIIII	CTRY	
99.9	KTCS	IIIIIIII	CTRY	
AM 1280	KPRV	I	CTRY	
AM 1410	KTCS	IIIII	CTRY	
106.3	KZKZ	III	RLG AC	
AM 950	KFSA	III	RLG	
AM 1230	KFPW	I	EZ	
AM 1320	KWHN	IIIII	NWS, TLK	

LITTLE ROCK, AR

98.5	KURB	IIIIIIII	HOT AC	
99.5	KYFX	III	AC	
102.1	KOKY	III	AC	
100.3	KQAR	III	TOP 40	
106.3	KHTE	III	TOP 40	
94.1	KKPT	IIIIIIII	CLS RK	
101.1	KDRE	III	DVRS	
105.1	KMJX	IIIIIIII	CLS RK	
107.7	KLAL	IIIII	DVRS	
94.9	KOLL	IIIIIIII	OLDS	
96.1	KSSN	IIIIIIII	CTRY	
96.5	KHUG	III	CTRY	
106.7	KDDK	IIIIIIII	CTRY	
88.3	KABF	IIIIIIII	DVRS	P
89.1	KUAR	IIIIIIII	NWS, NPR, PRI	P
90.5	KLRE	III	CLS, NPR, PRI	P
91.3	KUCA	III	NWS, TLK	C
AM 1090	KAAY	IIIIIIII	RLG AC	
AM 1150	KLRG	IIIII	GOSP	
AM 1250	KLIH	I	GOSP	
AM 1440	KITA	IIIII	GOSP	
92.3	KIPR	IIIIIIII	URBAN	
101.7	KKRN	I	NWS, TLK	
102.5	KARN	III	NWS, TLK	
103.7	KSYG	IIIIIIII	NWS, TLK	
AM 920	KARN	IIIIIIII	NWS, TLK	

CEDAR FALLS, IA

96.1	KCVM	III	AC	
99.3	KWAY	I	AC	

CEDAR FALLS • CEDAR RAPIDS • DES MOINES DUBUQUE • DAVENPORT, IA

	FREQ	STATION	POWER	FORMAT	
	107.9	KFMW	IIIIIII	MOD RK	
	105.7	KOKZ	IIIIIII	OLDS	
	98.5	KKCV	III	CTRY	
AM	1330	KWLO	IIIII	BB, NOS	
	88.1	KBBG	III	GOSP	P
	89.5	KHKE	III	JZZ, CLS, PRI	P
	90.9	KUNI	IIIIIII	DVRS, NPR, PRI	P
	101.9	KNWS	IIIIIII	RLG	C
AM	1090	KNWS	I	RLG	C
AM	1250	KCNZ	I	NWS, TLK	
AM	1540	KXEL	IIIIIII	NWS, TLK	

CEDAR RAPIDS, IA

	FREQ	STATION	POWER	FORMAT	
	96.5	WMT	IIIIIII	AC	
	104.5	KDAT	IIIIIII	AC	
	98.1	KHAK	IIIII	CTRY	
AM	1450	KMRY	I	BB, NOS	
	88.3	KCCK	III	JZZ, CLS, PRI	P
	102.9	KZIA	IIIII	TOP 40, NPR	
AM	1360	KTOF	I	RLG AC	
AM	600	WMT	IIIIIII	FS	
AM	1600	KCRG	IIIII	NWS	

DES MOINES, IA

	FREQ	STATION	POWER	FORMAT	
	100.3	KMXD	IIIIIII	HOT AC	
	102.5	KSTZ	IIIIIII	HOT AC	
	104.1	KLTI	IIIIIII	LGHT AC	
	106.3	KYSY	III	LGHT AC	
	105.1	KCCQ	III	TOP 40	
	94.9	KGGO	IIIIIII	CLS RK	
	107.5	KKDM	IIIII	DVRS	
	93.3	KIOA	IIIIIII	OLDS	
	92.5	KJJY	IIIIIII	CTRY	
	97.3	KHKI	IIIII	CTRY	
AM	1310	KDLS	I	CTRY	
AM	1350	KRNT	IIIII	BB, NOS	
	88.5	KURE	I	DVRS	P
	89.3	KUCB	III	DVRS	P
	90.1	WOI	IIIII	CLS	C
AM	640	WOI	IIIIIII	NWS, TLK, NPR	C
AM	1150	KWKY	I	RLG	
AM	940	KXTK	IIIIIII	NWS, TLK	
AM	1040	WHO	IIIIIII	NWS, TLK	

DUBUQUE, IA

	FREQ	STATION	POWER	FORMAT	
	92.9	KATF	IIIII	AC	
	105.3	KLYV	IIIII	TOP 40	
	97.3	KGRR	III	CLS RK	
	102.3	KXGE	III	CLS RK	
	107.1	WPVL	I	OLDS	
AM	1490	WDBQ	I	OLDS	
	97.7	WGLR	I	CTRY	
	99.3	KDST	III	CTRY	
	103.3	KIKR	III	CTRY	
	107.5	WJOD	III	CTRY	
AM	1590	WPVL	I	BB, NOS	
	90.5	WSUP	I	AC	C
AM	1370	KDTH	IIIII	NWS, TLK	

DAVENPORT, IA

	FREQ	STATION	POWER	FORMAT	
	93.9	WJRE	III	AC	
	98.9	WHTS	IIIII	HOT AC	
	93.5	KORB	III	DVRS	
	96.9	WXLP	IIIII	ROCK	

IA

77

	FREQ	STATION	POWER	FORMAT	
	106.5	KCQQ	|||	CLS RK	
	101.3	KUUL	|||||	OLDS	
	103.7	WLLR	|||||||	CTRY	
AM	1230	WLLR			CTRY	
AM	1270	WKBF	|||||	BB, NOS	
	88.5	KALA			DVRS	P
	90.3	WVIK	|||||	CLS, NPR, PRI	P
AM	1170	KJOC			SPRTS	
AM	1420	WOC	|||||	NWS, TLK	
AM	1450	WKEI			NWS, TLK	

SIOUX CITY, IA

	FREQ	STATION	POWER	FORMAT	
	107.1	KSFT	|||	LGHT AC	
	95.5	KGLI	|||||||	TOP 40	
	97.9	KSEZ	|||||||	CLS RK	
AM	620	KMNS	|||||	CTRY	
AM	1470	KWSL	|||||	BB, NOS	
	90.3	KWIT	|||||||	CLS, NPR, PRI	C
	103.3	KTFC	|||||||	GOSP	
AM	1360	KSCJ	|||||	NWS, TLK	

TOPEKA, KS

	FREQ	STATION	POWER	FORMAT	
	99.3	KWIC	|||	HOT AC	
	107.7	KMAJ	|||||||	AC	
	100.3	KDVV	|||||||	CLS RK	
	92.9	KANS	|||||||	OLDS	
	97.3	WIBW	|||||||	CTRY	
	106.9	KTPK	|||||||	CTRY	
AM	1490	KTOP			BB, NOS	
AM	580	WIBW	|||||||	NWS, TLK	
AM	1440	KMAJ	|||||	NWS, TLK	

WICHITA, KS

	FREQ	STATION	POWER	FORMAT	
	97.9	KRBB	|||||||	AC	
	98.7	KAYY	|||	HOT AC	
	93.9	KDGS	|||	TOP 40	
	107.3	KKRD	|||||||	TOP 40	
	95.1	KICT	|||||||	ROCK	
	96.3	KRZZ	|||||	CLS RK	
	104.5	KLLS	|||||||	70's RK	
	103.7	KEYN	|||||||	OLDS	
	101.3	KFDI	|||||||	CTRY	
AM	1070	KFDI	|||||	CTRY	
	105.3	KWSJ	|||||||	JZZ	
	88.1	KBCU			DVRS	C
	89.1	KMUW	|||||	NWS, TLK, NPR, PRI	P
	99.1	KTLI	|||||||	RLG AC	
	92.3	KOEZ	|||||||	EZ	
AM	950	KJRG	|||	RLG	
AM	1240	KNSS			NWS, TLK	
AM	1330	KFH	|||||	NWS, TLK	
AM	1480	KQAM	|||||	SPRTS	

ALEXANDRIA, LA

	FREQ	STATION	POWER	FORMAT	
	93.9	KFAD	|||	AC	
	93.1	KQID	|||||||	TOP 40	
	96.9	KZMZ	|||||||	CLS RK	
	92.1	KLIL	|||	OLDS	
	104.3	KEZP	|||||||	OLDS	
	97.7	KAPB	|||	CTRY	
	100.3	KRRV	|||||||	CTRY	
	103.5	KLAA	|||||||	CTRY	

IA
KS
LA

	FREQ	STATION	POWER	FORMAT	
	90.7	KLSA	IIIIIIII	CLS, NPR, PRI	P
AM	580	KLBG	IIIIIIII	GOSP	
	102.3	KBCE	III	URBAN	
AM	970	KSYL	III	NWS, TLK	

BATON ROUGE, LA

	FREQ	STATION	POWER	FORMAT	
	96.1	KRVE	IIIIIIII	LGHT AC	
	102.5	WLSS	IIIIIIII	TOP 40	
	98.1	WDGL	IIIIIIII	CLS RK	
	104.9	KKAY	III	OLDS	
	107.3	WTGE	III	OLDS	
	100.7	WXCT	IIIIIIII	CTRY	
	101.5	WYNK	IIIIIIII	CTRY	
	89.3	WRKF	IIIII	CLS, NPR, PRI	P
	91.1	KLSU	I	JZZ	C
AM	910	WNDC	III	GOSP	
AM	1260	KBRH	I	URBAN	
AM	1460	WXOK	IIIII	URBAN	
AM	1150	WJBO	IIIII	NWS, TLK	
AM	1210	WSKR	IIIII	SPRTS	
AM	1300	WIBR	IIIII	SPRTS	

LAFAYETTE, LA

	FREQ	STATION	POWER	FORMAT	
	99.9	KTDY	IIIIIIII	HOT AC	
	94.5	KSMB	IIIIIIII	TOP 40	
	96.5	KFTE	III	MOD RK	
	105.5	KJJB	III	OLDS	
	106.7	KLTW	III	OLDS	
	107.1	KOGM	I	OLDS	
AM	1240	KANE	I	OLDS	
	99.1	KXKC	IIIIIIII	CTRY	
AM	1230	KSLO	I	CTRY	
AM	1490	KEUN	I	CTRY	
AM	1360	KNIR	I	BB, NOS	
AM	1450	KSIG	IIIIIIII	BB, NOS	
	88.7	KRVS	IIIII	DVRS, NPR, PRI	P
	102.9	KAJN	IIIIIIII	RLG AC	
AM	1520	KDYS	IIIII	KIDS	
	93.7	KTBT	IIIIIIII	URBAN	
	95.5	KRRQ	III	URBAN	
	104.7	KNEK	III	URBAN	
	105.9	KVOL	III	SPRTS	
AM	770	KJCB	IIIII	URBAN	
AM	1330	KVOL	IIIII	SPRTS	
AM	1420	KPEL	I	NWS, TLK	

LAKE CHARLES, LA

	FREQ	STATION	POWER	FORMAT	
	99.5	KHLA	IIIIIIII	AC	
	103.7	KBIU	IIIII	AC	
	101.3	KKGB	III	CLS RK	
	96.1	KYKZ	IIIIIIII	CTRY	
	105.3	KZWA	IIIII	URBAN	
AM	1580	KXZZ	I	URBAN	
AM	1400	KAOK	I	NWS, TLK	
AM	1470	KLCL	IIIII	SPRTS	

MONROE, LA

	FREQ	STATION	POWER	FORMAT	
	101.9	KNOE	IIIIIIII	TOP 40	
	105.3	KLIP	IIIII	CLS RK	
	104.1	KJLO	IIIIIIII	CTRY	
	106.1	KMYY	IIIIIIII	CTRY	
	540	KNOE	IIIIIIII	CTRY	
	90.3	KEDM	IIIIIIII	CLS, NPR, PRI	P
	91.1	KNLU	III	AC	C

LA

LA MN

FREQ		STATION	POWER	FORMAT	
	100.9	KHLL	III	RLG AC	
AM	1230	KLIC	I	RLG AC	
	98.3	KYEA	IIIII	URBAN	
AM	1440	KMLB	IIIII	NWS, TLK	

NEW ORLEANS, LA

	FREQ	STATION	POWER	FORMAT	
	97.1	WEZB	IIIIIIII	HOT AC	
	101.9	WLMG	IIIIIIII	AC	
	105.3	WLTS	IIIIIIII	HOT AC	
	92.3	WCKW	IIIIIIII	ROCK	
	99.5	WRNO	IIIIIIII	CLS RK	
	95.7	WTKL	IIIIIIII	OLDS	
	94.7	WYLA	III	CTRY	
	101.1	WNOE	IIIIIIII	CTRY	
	104.7	WYLK	III	CTRY	
AM	830	WFNO	IIIIIIII	SPAN	
AM	1280	WODT	IIIII	BLK, R&B	
AM	1450	WBYU	I	BB, NOS	
	88.3	WRBH	III	NWS, TLK	P
	89.9	WWNO	IIIIIIII	CLS, NPR, PRI	P
	90.7	WWOZ	III	DVRS	P
AM	1230	WBOK	I	GOSP	
	94.9	WADU	III	EZ	
AM	800	WSHO	III	RLG	
AM	940	WYLD	IIIIIIII	RLG	
AM	1060	WLNO	IIIIIIII	RLG	
	93.3	WQUE	IIIIIIII	URBAN	
	98.5	WYLD	IIIII	URBAN	
AM	690	WTIX	IIIIIIII	NWS, TLK	
AM	870	WWL	IIIIIIII	NWS, TLK	
AM	990	WGSO	III	NWS	
AM	1350	WSMB	IIIII	NWS, TLK	

SHREVEPORT, LA

	FREQ	STATION	POWER	FORMAT	
	96.5	KVKI	IIIIIIII	AC	
	94.5	KRUF	IIIIIIII	TOP 40	
	92.1	KLKL	III	OLDS	
AM	1240	KASO	I	OLDS	
	93.7	KITT	IIIIIIII	CTRY	
	101.1	KRMD	IIIIIIII	CTRY	
AM	1320	KNCB	IIIII	CTRY	
	89.9	KDAQ	IIIIIIII	CLS, NPR, PRI	P
	91.3	KSCL	I	DVRS	C
	99.7	KMJJ	IIIII	URBAN	
	103.7	KDKS	III	URBAN	
AM	710	KEEL	IIIIIIII	NWS, TLK	
AM	1130	KWKH	IIIIIIII	SPRTS	
AM	1340	KRMD	I	SPRTS	

DULUTH, MN

	FREQ	STATION	POWER	FORMAT	
	95.7	KDAL	IIIIIIII	LGHT AC	
AM	610	KDAL	IIIIIIII	LGHT AC	
	92.1	WWAX	III	MOD RK	
	94.9	KQDS	IIIIIIII	ROCK	
	102.5	KRBR	IIIIIIII	MOD RK	
	107.7	KUSZ	III	CLS RK	
	101.7	KLDJ	III	OLDS	
	98.9	KTCO	IIIIIIII	CTRY	
	105.1	KKCB	IIIIIIII	CTRY	
	89.9	WHSA	IIIII	CLS, NPR, PRI	P
	91.3	KUWS	IIIIIIII	NWS, TLK, NPR, PRI	P
	92.9	WSCD	IIIIIIII	CLS, NPR, PRI	P

FREQ	STATION	POWER	FORMAT	
90.5	KDNI	I	RLG	C
97.3	KDNW	IIIII	RLG	C
AM 560	WEBC	IIIIIIII	NWS, TLK	
AM 710	WDSM	IIIIIIII	SPRTS	

MINN.- ST. PAUL, MN

FREQ	STATION	POWER	FORMAT	
94.5	KSTP	IIIIIIII	HOT AC	
102.9	WLTE	IIIIIIII	AC	
101.3	KDWB	III	TOP 40	
92.5	KQRS	IIIIIIII	CLS RK	
93.7	KXXR	IIIIIIII	ROCK	
97.1	KTCZ	IIIIIIII	ALT RK	
100.3	WRQC	IIIIIIII	CLS RK	
105.1	KZNR	III	DVRS	
105.3	KZNT	III	DVRS	
105.7	KZNZ	I	DVRS	
107.9	KQQL	IIIIIIII	OLDS	
102.1	KEEY	IIIIIIII	CTRY	
107.1	WIXK	III	CTRY	
AM 1590	WIXK	IIIII	CTRY	
104.1	KMJZ	IIIIIIII	JZZ	
AM 1220	WEZU	IIIII	BB, NOS	
AM 1400	KLBB	I	BB, NOS	
88.7	WRFW	I	DVRS	C
89.9	KMOJ	I	URBAN	P
90.3	KFAI	III	DVRS	P
91.1	KNOW	IIIIIIII	NWS, TLK, NPR, PRI	P
99.5	KSJN	IIIII	CLS, NPR, PRI	P
AM 770	KUOM	IIIIIIII	DVRS	P
AM 980	KKMS	IIIIIIII	RLG AC	
AM 1280	WWTC	IIIII	DVRS	
AM 830	WCCO	IIIIIIII	FS	
AM 900	KTIS	IIIIIIII	RLG AC	C
AM 1130	KFAN	IIIIIIII	SPRTS	
AM 1330	WMNN	IIIII	NWS	
AM 1500	KSTP	IIIIIIII	NWS, TLK	

ROCHESTER, MN

FREQ	STATION	POWER	FORMAT	
105.3	KYBA	IIIIIIII	LGHT AC	
106.9	KROC	IIIIIIII	TOP 40	
101.7	KRCH	IIIII	CLS RK	
96.5	KWWK	IIIIIIII	CTRY	
90.7	KZSE	III	NWS, TLK, NPR, PRI	P
91.7	KLSE	IIIIIIII	CLS, NPR, PRI	P
89.9	KRPR	III	RLG	C
97.5	KNXR	IIIIIIII	EZ	
AM 1270	KWEB	IIIII	SPRTS	
AM 1340	KROC	I	NWS, TLK	

ST. CLOUD, MN

FREQ	STATION	POWER	FORMAT	
96.7	KKSR	IIIIIIII	AC	
104.7	KCLD	IIIIIIII	AC	
101.7	WHMH	III	ROCK	
103.7	KLZZ	III	CLS RK	
94.9	KMXK	IIIII	OLDS	
98.1	WWJO	IIIIIIII	CTRY	
98.9	KZPK	IIIII	CTRY	
105.5	KDDG	III	CTRY	
AM 1390	KXSS	III	BB, NOS	
88.1	KVSC	III	AC	C
88.9	KNSR	IIIIIIII	NWS, TLK, NPR, PRI	P
90.1	KSJR	IIIIIIII	CLS, NPR, PRI	P

	FREQ	STATION	POWER	FORMAT	
	92.9	KKJM	IIIII	RLG AC	
AM	1240	WJON	I	NWS, TLK	
AM	1450	KNSI	I	NWS, TLK	

COLUMBIA, MO

	FREQ	STATION	POWER	FORMAT	
	101.5	KPLA	IIIIIIII	AC	
	96.7	KCMQ	IIIII	CLS RK	
	98.3	KFMZ	IIIIIIII	DVRS	
	102.3	KBXR	IIIII	ALT RK	
	106.1	KOQL	IIIIIIII	OLDS	
	88.1	KCOU	I	DVRS	C
	89.5	KOPN	III	NWS, TLK, NPR, PRI	P
	90.5	KWWC	I	JZZ	C
	91.3	KBIA	IIIIIIII	CLS, NPR, PRI	P
	92.1	KMFC	IIIII	RLG AC	
AM	1400	KFRU	I	NWS, TLK	

JOPLIN, MO

	FREQ	STATION	POWER	FORMAT	
	93.9	KJMK	IIIIIIII	AC	
	95.1	KMXL	IIIII	LGHT AC	
	92.5	KSYN	IIIII	TOP 40	
	97.9	KXDG	III	CLS RK	
	103.5	KWXD	III	OLDS	
	99.7	KBTN	I	CTRY	
	102.5	KIXQ	IIIII	CTRY	
AM	1420	KBTN	I	CTRY	
AM	1490	KDMO	I	CTRY	
AM	1230	KWAS	I	BB, NOS	
	88.7	KXMS	III	CLS	P
AM	1450	WMBH	I	SPRTS	
AM	1560	KQYX	IIIII	NWS, TLK	

KANSAS CITY, MO

	FREQ	STATION	POWER	FORMAT	
	98.9	KQRC	IIIIIIII	ROCK	
	99.7	KYYS	IIIIIIII	CLS RK	
	101.1	KCFX	IIIIIIII	CLS RK	
	102.1	KOZN	IIIIIIII	DVRS	
	107.3	KNRX	IIIIIIII	DVRS	
	94.9	KCMO	IIIIIIII	OLDS	
	94.1	KFKF	IIIIIIII	CTRY	
	104.3	KBEQ	IIIIIIII	CTRY	
AM	610	WDAF	IIIIIIII	CTRY	
AM	810	WHB	IIIIIIII	CTRY	
AM	1030	KOWW	I	CTRY	
	96.5	KXTR	IIIIIIII	CLS	
	106.5	KCIY	IIIIIIII	JZZ	
AM	1340	KFEZ	I	BB, NOS	
	89.3	KCUR	IIIIIIII	NWS, TLK, NPR, PRI	P
	91.9	KWJC	III	RLG AC	C
	92.3	KCCV	III	RLG	
	103.3	KPRS	IIIIIIII	URBAN	
AM	710	KCMO	IIIIIIII	NWS, TLK	
AM	980	KMBZ	IIIIIIII	NWS, TLK	
AM	1190	KPHN	IIIII	NWS, TLK	

SPRINGFIELD, MO

	FREQ	STATION	POWER	FORMAT	
	101.3	KTXR	IIIIIIII	AC	
	104.1	KZRQ	III	HOT AC	
	105.9	KGBX	IIIIIIII	AC	
	95.5	KTOZ	IIIII	MOD RK	
	97.3	KXUS	IIIII	ROCK	
	98.7	KWTO	IIIIIIII	CLS RK	
	104.7	KKLH	III	CLS RK	
	105.1	KOSP	III	OLDS	

82

	FREQ	STATION	POWER	FORMAT	
	94.7	KTTS	|||||||	CTRY	
	96.5	KLTQ	III	CTRY	
AM	1260	KTTS	|||||	CTRY	
AM	1400	KGMY	|||||	BB, NOS	
	91.1	KSMU	III	NWS, TLK, NPR, PRI	P
	99.5	KADI	III	RLG AC	
AM	560	KWTO	|||||||	NWS, TLK	

ST. LOUIS, MO

	FREQ	STATION	POWER	FORMAT	
	98.1	KYKY	|||||||	AC	
	98.7	WJKK	|||||||	AC	
	101.1	WVRV	|||||||	AC	
	102.5	KEZK	|||||||	LGHT AC	
	107.7	KSLZ	|||||||	TOP 40	
	93.7	KSD	|||||	CLS RK	
	94.7	KSHE	|||||||	ROCK	
	96.3	KIHT	|||||||	70's RK	
	97.1	KXOK	|||||||	MOD RK	
	104.1	WXTM	|||||||	ROCK	
	106.7	WSTZ	|||||||	CLS RK	
	103.3	KLOU	|||||||	OLDS	
	92.3	WIL	|||||	CTRY	
	99.9	KFAV	|||||||	CTRY	
	106.5	WKKX	|||||||	CTRY	
AM	730	KWRE	|||||	CTRY	
	100.3	KATZ	|||||	BLK, R&B	
	104.9	KMJM	III	BLK, R&B	
	99.1	KFUO	|||||||	CLS	
AM	1430	WRTH	|||||	BB, NOS	
	88.1	KDHX	|||||	DVRS, NPR	P
	88.7	WSIE	|||||	JZZ, NPR, PRI	P
	89.1	KCLC	III	AC	C
	89.5	KCFV	I	DVRS	C
	89.7	KYMC	I	CLS RK	C
	89.9	WLCA	I	DVRS	C
	90.7	KWMU	|||||||	NWS, TLK, NPR, PRI	P
AM	1460	KIRL	|||||	GOSP	
AM	1600	KATZ	|||||	GOSP	
AM	630	KJSL	|||||||	RLG	
AM	1260	WIBV	|||||	KIDS	
AM	1320	KSIV	|||||	RLG	
AM	1490	WESL	I	URBAN	
AM	550	KTRS	|||||||	NWS, TLK	
AM	590	KFNS	|||||	SPRTS	
AM	920	WGNU	III	NWS, TLK	
AM	1120	KMOX	|||||||	NWS, TLK	
AM	1220	KLPW	I	NWS, TLK	

BISMARCK, ND

	FREQ	STATION	POWER	FORMAT	
	92.9	KYYY	|||||||	HOT AC	
AM	550	KFYR	|||||||	AC	
	96.5	KBYZ	|||||||	CLS RK	
	101.5	KSSS	|||||||	CLS RK	
	98.7	KACL	|||||||	OLDS	
	94.5	KQDY	|||||||	CTRY	
	97.5	KKCT	|||||||	CTRY	
AM	1130	KBMR	|||||||	CTRY	
AM	1270	KLXX	I	BB, NOS	
	90.5	KCND	|||||||	NWS, TLK, NPR, PRI	P
	104.7	KNDR	|||||||	RLG	

	FREQ	STATION	POWER	FORMAT	
	93.7	WDAY	IIIIIII	TOP 40	
	98.7	KQWB	IIIIIII	ROCK	
	107.9	KPFX	IIIIIII	CLS RK	
	92.7	KPHT	III	OLDS	
	96.7	KOCL	III	OLDS	
	99.9	KVOX	IIIII	CTRY	
	101.9	KFGO	IIIIIII	CTRY	
AM	1550	KQWB	IIIII	BB, NOS	
	90.3	KCCD	III	NWS, TLK, NPR, PRI	P
	91.1	KCCM	IIIIIII	CLS, NPR, PRI	P
	91.9	KDSU	IIIIIII	JZZ, NPR	P
	97.9	KFNW	IIIIIII	RLG AC	C
AM	1200	KFNW	IIIII	RLG	C
AM	790	KFGO	IIIIIII	NWS, TLK	
AM	970	WDAY	IIIIIII	NWS, TLK	
AM	1280	KVOX	IIIII	SPRTS	

		GRAND FORKS, ND			
	92.9	KKXL	IIIII	AC	
	96.1	KQHT	IIIII	LGHT AC	
	104.3	KZLT	IIIII	AC	
AM	1260	KROX	I	AC	
	107.5	KJKJ	IIIII	CLS RK	
	94.7	KNOX	IIIII	CTRY	
	97.1	KYCK	IIIII	CTRY	
AM	1440	KKXL	I	BB, NOS	
	89.3	KUND	III	CLS, NPR, PRI	P
	90.7	KFJM	I	ALT RK, NPR	P
AM	1370	KUND	I	NWS, TLK, NPR, PRI	
	107.1	KKEQ	IIIIIII	RLG AC	
AM	1480	KKCQ	IIIII	RLG	
AM	1310	KNOX	IIIII	NWS, TLK	
AM	1590	KCNN	IIIII	NWS, TLK	

		LINCOLN, NE			
	107.3	KEZG	IIIIIII	AC	
AM	1240	KFOR	I	AC	
	102.7	KFRX	IIIII	TOP 40	
	95.1	KRKR	IIIII	CLS RK	
	106.3	KIBZ	I	MOD RK	
	105.3	KKUL	I	OLDS	
	89.3	KZUM	I	DVRS	P
	90.3	KRNU	I	AC	C
	90.9	KUCV	III	CLS, NPR, PRI	P
AM	1400	KLIN	I	NWS, TLK	
AM	1480	KLMS	IIIII	SPRTS	

		OMAHA, NE			
	96.1	KEFM	IIIIIII	AC	
	104.5	KSRZ	IIIIIII	HOT AC	
	98.5	KQKQ	IIIIIII	TOP 40	
	92.3	KEZO	IIIIIII	ROCK	
	93.3	KTNP	III	MOD RK	
	105.9	KKCD	IIIII	CLS RK	
	99.9	KGOR	IIIIIII	OLDS	
	94.1	WOW	IIIII	CTRY	
	101.5	KISP	III	CTRY	
AM	590	WOW	IIIIIII	CTRY	
AM	1420	KBBX	I	SPAN	
	88.9	KNOS	III	BLK, R&B	P
	89.7	KIWR	IIIIIII	ALT RK	C
	90.7	KVNO	III	CLS	P

	FREQ	STATION	POWER	FORMAT	
AM	660	KCRO	IIIII	RLG	
AM	1110	KFAB	IIIIIIII	NWS, TLK	
AM	1180	KOIL	IIIII	SPRTS	
AM	1290	KKAR	IIIII	NWS, TLK	
AM	1490	KOSR	I	SPRTS	

LAWTON, OK

	FREQ	STATION	POWER	FORMAT	
	99.5	KBZQ	III	LGHT AC	
	95.3	KMGZ	III	TOP 40	
	94.1	KZCD	III	CLS RK	
	107.3	KVRW	IIIIIIII	OLDS	
	101.5	KLAW	IIIIIIII	CTRY	
	98.1	KJMZ	I	URBAN	
AM	1380	KXCA	I	SPRTS	

OKLAHOMA CITY, OK

	FREQ	STATION	POWER	FORMAT	
	94.7	KQSR	IIIIIIII	AC	
	98.9	KYIS	IIIIIIII	HOT AC	
	104.1	KMGL	IIIIIIII	AC	
AM	1340	KEBC	I	AC	
	102.7	KJYO	IIIIIIII	TOP 40	
	100.5	KATT	IIIIIIII	ROCK	
	107.7	KRXO	IIIIIIII	CLS RK	
AM	1450	KGFF	I	70's RK	
	92.5	KOMA	IIIIIIII	OLDS	
AM	1520	KOMA	IIIIIIII	OLDS	
	96.1	KXXY	IIIIIIII	CTRY	
	101.9	KTST	IIIIIIII	CTRY	
	97.9	KTNT	III	JZZ	
	88.9	KOCC	III	AC	C
	90.1	KCSC	IIIII	CLS, NPR, PRI	P
	105.7	KROU	I	NWS, NPR, PRI	P
	106.3	KGOU	III	NWS, NPR, PRI	P
AM	800	KQCV	IIIII	RLG	
	104.9	KNTL	III	SPRTS	
AM	640	WWLS	IIIII	SPRTS	
AM	930	WKY	IIIIIIII	NWS, TLK	
AM	1000	KTOK	IIIIIIII	NWS, TLK	
AM	1400	KNOR	I	NWS, TLK	

TULSA, OK

	FREQ	STATION	POWER	FORMAT	
	92.9	KBEZ	IIIIIIII	AC	
	96.5	KRAV	IIIIIIII	HOT AC	
	97.5	KMOD	IIIIIIII	CLS RK	
	103.3	KJSR	IIIIIIII	70's RK	
	104.5	KMYZ	IIIIIIII	MOD RK	
	106.1	KQLL	IIIIIIII	OLDS	
	95.5	KWEN	IIIIIIII	CTRY	
	98.5	KVOO	IIIIIIII	CTRY	
	102.3	KTFX	III	CTRY	
	104.9	KREK	III	CTRY	
AM	1170	KVOO	IIIIIIII	CTRY	
	92.1	KOAS	IIIIIIII	JZZ	
	89.5	KWGS	IIIIIIII	NWS, TLK, NPR, PRI	P
	91.3	KRSC	III	HOT AC	C
	94.5	KEMX	III	RLG AC	
	100.9	KXOJ	III	RLG AC	
AM	1340	KTOW	I	GOSP	
AM	970	KCFO	IIIII	RLG	
	105.3	KJMM	III	URBAN	
AM	740	KRMG	IIIIIIII	NWS, TLK	
AM	1300	KAKC	IIIII	SPRTS	

NE
OK

85

	FREQ	STATION	POWER	FORMAT	
AM	1430	KQLL	IIIII	SPRTS	

RAPID CITY, SD

	FREQ	STATION	POWER	FORMAT	
	93.1	KRCS	IIIIIIII	AC	
	93.9	KKMK	IIIIIIII	AC	
	100.3	KFXS	IIIII	CLS RK	
AM	920	KKLS	IIIIIIII	OLDS	
	98.7	KOUT	IIIII	CTRY	
	104.1	KIQK	IIIIIIII	CTRY	
AM	1150	KIMM	IIIII	CTRY	
AM	1340	KTOQ	I	BB, NOS	
	89.3	KBHE	III	CLS, NPR, PRI	P
	97.1	KPSD	IIIIIIII	CLS, NPR, PRI	P
	97.9	KLMP	IIIII	RLG	
AM	1380	KOTA	IIIII	NWS, TLK	

SIOUX FALLS, SD

	FREQ	STATION	POWER	FORMAT	
	92.5	KELO	IIIIIIII	LGHT AC	
	97.3	KMXC	III	HOT AC	
	104.7	KKLS	IIIIIIII	TOP 40	
	102.7	KYBB	IIIII	CLS RK	
	103.7	KRRO	III	ROCK	
AM	1320	KELO	IIIII	OLDS	
	101.9	KTWB	IIIIIIII	CTRY	
	88.1	KRSD	I	CLS, NPR, PRI	P
	89.1	KAUR	I	ROCK	C
	90.9	KCSD	I	CLS, NPR, PRI	P
	94.5	KCFS	I	RLG	C
AM	1140	KSOO	IIIII	NWS, TLK	
AM	1230	KWSN	I	NWS, TLK	

ABILENE, TX

	FREQ	STATION	POWER	FORMAT	
	100.7	KORQ	IIIIIIII	AC	
	102.7	KHXS	IIIIIIII	AC	
	103.7	KCDD	IIIIIIII	TOP 40	
	106.3	KFQX	I	CLS RK	
	107.9	KEYJ	IIIIIIII	ROCK	
	99.7	KBCY	IIIIIIII	CTRY	
	105.1	KEAN	IIIIIIII	CTRY	
	98.1	KKHR	IIIII	SPAN	
AM	1470	KBBA	IIIII	SPAN	
	89.7	KACU	III	AC, NPR	C
AM	1280	KGMM	I	SPRTS	
AM	1340	KYYD	I	NWS, TLK	

AMARILLO, TX

	FREQ	STATION	POWER	FORMAT	
	105.7	KAEZ	I	AC	
	93.1	KQIZ	IIIIIIII	TOP 40	
	98.7	KPRF	IIIII	TOP 40	
	94.1	KFNX	IIIIIIII	ROCK	
	99.7	KBZD	III	ALT RK	
	107.9	KZRK	IIIIIIII	ROCK	
	107.1	KPUR	III	OLDS	
	96.9	KMML	IIIIIIII	CTRY	
	97.9	KGNC	IIIIIIII	CTRY	
	101.9	KATP	IIIIIIII	CTRY	
AM	1360	KDJW	I	CTRY	
AM	1310	KZIP	I	SPAN	
AM	940	KIXZ	IIIIIIII	BB, NOS	
	89.9	KACV	IIIIIIII	DVRS	C
	91.1	KWTS	I	TOP 40	C
AM	710	KGNC	IIIIIIII	NWS, TLK	

OK
SD
TX

TX

	FREQ	STATION	POWER	FORMAT	
AM	1440	KPUR	IIIII	SPRTS	

AUSTIN, TX

	FREQ	STATION	POWER	FORMAT	
	94.7	KAMX	IIIIIIII	HOT AC	
	95.5	KKMJ	IIIIIIII	AC	
	96.7	KHFI	IIIIIIII	TOP 40	
	104.3	KQBT	IIIII	TOP 40	
	93.7	KLBJ	IIIIIIII	MOD RK	
	102.3	KPEZ	IIIIIIII	CLS RK	
	107.1	KGSR	IIIIIIII	ALT RK	
	107.7	KAHK	III	CLS RK	
	103.5	KEYI	IIIIIIII	OLDS	
	98.1	KVET	IIIIIIII	CTRY	
	100.7	KASE	IIIIIIII	CTRY	
	92.1	KQQQ	III	SPAN	
	92.5	KKLB	III	SPAN	
AM	1260	KTAE	I	SPAN	
AM	1440	KELG	I	SPAN	
AM	1560	KTXZ	III	SPAN	
	88.7	KAZI	III	BLK, R&B	P
	89.5	KMFA	IIIIIIII	CLS	P
	90.5	KUT	IIIIIIII	DVRS, NPR	P
	91.7	KVRX	I	DVRS	C
AM	970	KIXL	III	RLG	
AM	1370	KJCE	IIIII	URBAN	
AM	590	KLBJ	IIIIIIII	NWS, TLK	
AM	1300	KVET	IIIII	NWS, TLK	
AM	1490	KFON	I	SPRTS	

BEAUMONT, TX

	FREQ	STATION	POWER	FORMAT	
	104.5	KKMY	IIIII	AC	
	94.1	KQXY	IIIIIIII	TOP 40	
	106.1	KIOC	IIIIIIII	MOD RK	
	95.1	KYKR	IIIII	CTRY	
	97.5	KAYD	IIIII	CTRY	
AM	1450	KAYD	I	CTRY	
AM	1600	KOGT	I	CTRY	
AM	1250	KALO	IIIII	BLK, R&B	
	91.3	KVLU	III	CLS, NPR, PRI	P
	92.5	KTFA	IIIII	RLG AC	
	102.5	KTCX	IIIIIIII	URBAN	
AM	560	KLVI	IIIIIIII	NWS, TLK	
AM	1340	KOLE	I	NWS, TLK	

BRYAN/COLL. STATION, TX

	FREQ	STATION	POWER	FORMAT	
	104.7	KKYS	IIIII	AC	
	92.1	KTSR	III	CLS RK	
AM	1240	KTAM	I	DVRS	
	98.3	KORA	III	CTRY	
	99.5	KBMA	III	SPAN	
	89.1	KEOS	III	ALT RK, PRI	P
	90.9	KAMU	III	CLS, NPR, PRI	P
AM	1150	WTAW	I	NWS, TLK	

CORPUS CHRISTI, TX

	FREQ	STATION	POWER	FORMAT
	93.9	KMXR	IIIIIIII	AC
	95.5	KZFM	IIIIIIII	TOP 40
	94.7	KBSO	III	CLS RK
	101.3	KNCN	IIIII	MOD RK
	105.5	KRAD	III	MOD RK
	96.5	KLTG	IIIIIIII	OLDS
	107.3	KCCG	III	OLDS
	99.1	KRYS	IIIIIIII	CTRY
	103.7	KOUL	IIIIIIII	CTRY

	FREQ	STATION	POWER	FORMAT	
	98.3	KLHB	‖‖‖	SPAN	
	99.9	KSAB	‖‖‖	SPAN	
	105.1	KMIQ	‖‖	SPAN	
AM	1400	KUNO	‖	SPAN	
AM	1590	KDAE	‖	BB, NOS	
	90.3	KEDT	‖‖‖‖	CLS, NPR, PRI	P
	91.7	KBNJ	‖‖	RLG	P
AM	1360	KRYS	‖	KIDS	
AM	1150	KCCT	‖	SPRTS	
AM	1230	KSIX	‖	NWS	
AM	1440	KEYS	‖	NWS, TLK	

DALLAS / FT. WORTH, TX

	FREQ	STATION	POWER	FORMAT	
	97.9	KBFB	‖‖‖‖	AC	
	102.9	KDMX	‖‖‖‖	AC	
	103.7	KVIL	‖‖‖‖	AC	
	106.1	KHKS	‖‖‖‖	TOP 40	
	92.5	KZPS	‖‖‖‖	CLS RK	
	93.3	KKZN	‖‖‖	ALT RK	
	97.1	KEGL	‖‖‖‖	ROCK	
	98.7	KLUV	‖‖‖‖	OLDS	
	102.1	KTXQ	‖‖‖‖	OLDS	
AM	1190	KLUV	‖‖‖‖	OLDS	
	96.3	KSCS	‖‖‖‖	CTRY	
	99.5	KPLX	‖‖‖‖	CTRY	
	105.3	KYNG	‖‖‖‖	CTRY	
	89.3	KNON	‖‖‖‖	SPAN	P
	99.1	KHCK	‖‖‖‖	SPAN	
	106.7	KDXT	‖‖‖‖	SPAN	
	106.9	KZDF	‖‖	SPAN	
	107.1	KZDL	‖‖	SPAN	
AM	1270	KESS	‖‖‖	SPAN	
AM	1480	KDXX	‖‖‖	SPAN	
AM	1600	KRVA	‖‖‖	SPAN	
AM	730	KKDA	‖‖‖	BLK, R&B	
	101.1	WRR	‖‖‖‖	CLS	
	107.5	KOAI	‖‖‖‖	JZZ	
	88.1	KNTU	‖‖‖	JZZ	C
	88.7	KTCU	‖‖	CLS	C
	89.1	KMQX	‖	BB, NOS	P
	89.5	KYQX	‖	BB, NOS	P
	90.1	KERA	‖‖‖‖	NWS, TLK, NPR, PRI	P
	94.1	KLTY	‖‖‖‖	RLG AC	
AM	970	KHVN	‖‖	GOSP	
	88.3	KJCR	‖	RLG	C
	94.9	KWRD	‖‖‖‖	RLG	
	100.3	KRBV	‖‖‖‖	URBAN	
	104.5	KKDA	‖‖‖‖	URBAN	
AM	570	KLIF	‖‖‖‖	NWS, TLK	
AM	820	WBAP	‖‖‖‖	NWS, TLK	
AM	1080	KRLD	‖‖‖‖	NWS, TLK	
AM	1310	KTCK	‖‖‖	SPRTS	

EL PASO, TX

	FREQ	STATION	POWER	FORMAT	
	93.1	KSII	‖‖‖‖	AC	
	99.9	KTSM	‖‖‖‖	LGHT AC	
	102.1	KPRR	‖‖‖‖	TOP 40	
	95.5	KLAQ	‖‖‖‖	ROCK	
	92.3	KOFX	‖‖‖‖	OLDS	
	94.7	KATH	‖‖‖‖	CTRY	
	96.3	KHEY	‖‖‖‖	CTRY	
	93.9	KINT	‖‖‖‖	SPAN	
	97.5	KBNA	‖‖‖‖	SPAN	
	103.1	KPAS	‖‖	SPAN	

TX

FREQ	STATION	POWER	FORMAT	
AM 750	KAMA	IIIIIII	SPAN	
AM 920	KBNA	III	SPAN	
AM 1150	KSVE	IIIII	SPAN	
AM 1340	KVIV	I	SPAN	
88.5	KTEP	IIIIIII	CLS, NPR, PRI	P
89.5	KXCR	III	JZZ, CLS, PRI	P
AM 1590	KELP	IIIII	RLG	
AM 600	KROD	IIIIIII	NWS, TLK	
AM 690	KHEY	IIIIIII	SPRTS	
AM 1060	KFNA	IIIII	NWS, TLK	
AM 1380	KTSM	IIIII	NWS, TLK	

HOUSTON, TX

FREQ	STATION	POWER	FORMAT	
96.5	KHMX	IIIIIII	AC	
99.1	KODA	IIIIIII	AC	
99.9	KSHN	IIIIIII	AC	
103.3	KJOJ	IIIIIII	TOP 40	
104.1	KRBE	IIIIIII	TOP 40	
93.7	KKRW	IIIIIII	CLS RK	
101.1	KLOL	IIIIIII	ROCK	
107.5	KTBZ	IIIIIII	MOD RK	
94.5	KLDE	IIIIIII	OLDS	
92.9	KKBQ	IIIIIII	CTRY	
95.7	KIKK	IIIIIII	CTRY	
100.3	KILT	IIIIIII	CTRY	
100.7	KRTX	IIIII	SPAN	
102.9	KLTN	IIIIIII	SPAN	
106.5	KQQK	IIIIIII	SPAN	
AM 1010	KLAT	IIIII	SPAN	
AM 1320	KXYZ	IIIII	SPAN	
92.1	KRTS	IIIIIII	CLS	
AM 790	KBME	IIIIIII	BB, NOS	
88.7	KUHF	IIIIIII	CLS, NPR, PRI	P
89.7	KACC	I	CLS RK	C
90.1	KPFT	IIIII	DVRS, PRI	P
91.3	KPVU	III	JZZ, NPR	P
91.7	KTRU	IIIII	DVRS	C
104.9	KOVA	III	DVRS	
AM 1360	KWWJ	IIIII	GOSP	
AM 1590	KYOK	IIIII	GOSP	
90.9	KTSU	III	EZ	C
106.9	KKHT	IIIIIII	RLG	
97.1	KKTL	IIIIIII	NWS, TLK	
97.9	KBXX	IIIIIII	URBAN	
102.1	KMJQ	IIIIIII	URBAN	
AM 610	KILT	IIIIIII	SPRTS	
AM 700	KSEV	IIIIIII	NWS, TLK	
AM 740	KTRH	IIIIIII	NWS, TLK	
AM 950	KPRC	IIIIIII	NWS, TLK	
AM 1340	KKAM	I	SPRTS	

KILLEEN, TX

FREQ	STATION	POWER	FORMAT	
106.3	KOOC	III	AC	
98.3	KRYL	III	CLS RK	
107.3	KLFX	III	MOD RK	
105.5	KYUL	III	OLDS	
103.1	KOOV	IIIII	CTRY	
101.7	KLTD	III	RLG AC	
91.3	KNCT	IIIIIII	EZ	P
92.3	KIIZ	III	URBAN	
AM 1400	KTEM	III	NWS, TLK	

LAREDO, TX

FREQ	STATION	POWER	FORMAT	
98.1	KRRG	IIIIIII	TOP 40	

TX

89

TX

	FREQ	STATION	POWER	FORMAT	
	100.5	KBDR	IIIII	OLDS	
	94.9	KOYE	IIIIIIII	CTRY	
	92.7	KJBZ	III	SPAN	
	106.1	KNEX	III	SPAN	
AM	1300	KLAR	I	SPAN	
AM	1490	KLNT	I	SPAN	
	88.1	KHOY	III	EZ	P

LONGVIEW, TX

	FREQ	STATION	POWER	FORMAT	
	93.1	KTYL	IIIII	AC	
	106.5	KOOI	IIIIIIII	AC	
	107.3	KISX	IIIII	TOP 40	
	96.1	KKTX	IIIIIIII	CLS RK	
	97.7	KWRW	III	OLDS	
	101.5	KNUE	IIIIIIII	CTRY	
	104.1	KKUS	IIIII	CTRY	
	105.7	KYKX	IIIIIIII	CTRY	
AM	1400	KEBE	I	CTRY	
	92.1	KDOK	III	BB, NOS	
	88.7	KTPB	IIIIIIII	CLS, NPR, PRI	P
	102.3	KLJT	III	RLG AC	
	90.3	KBJS	IIIII	RLG	P
AM	690	KZEY	IIIII	URBAN	
AM	600	KTBB	IIIIIIII	NWS, TLK	
AM	1370	KFRO	I	NWS, TLK	
AM	1430	KEES	IIIII	NWS, TLK	
AM	1490	KYZS	I	NWS, TLK	

LUBBOCK, TX

	FREQ	STATION	POWER	FORMAT	
	101.1	KONE	IIIIIIII	LGHT AC	
	102.5	KZII	IIIIIIII	TOP 40	
	94.5	KFMX	IIIIIIII	ROCK	
	99.5	KCRM	IIIIIIII	CLS RK	
AM	950	KXTQ	IIIIIIII	CLS RK	
AM	1590	KDAV	I	OLDS	
	96.3	KLLL	IIIIIIII	CTRY	
	105.7	KRBL	III	CTRY	
	93.7	KXTQ	IIIIIIII	SPAN	
	106.5	KEJS	IIIII	SPAN	
AM	1420	KLFB	I	SPAN	
AM	1460	KBZO	I	SPAN	
	88.1	KTXT	III	AC	C
	89.1	KOHM	IIIII	CLS, NPR, PRI	P
	92.7	KJAK	IIIIIIII	RLG	
AM	580	KRFE	IIIII	EZ	
AM	790	KFYO	IIIIIIII	NWS, TLK	

BROWNSVILLE, TX

	FREQ	STATION	POWER	FORMAT	
	107.9	KVLY	IIIIIIII	AC	
	104.1	KBFM	IIIIIIII	TOP 40	
	92.7	KESO	III	DVRS	
	94.5	KFRQ	IIIIIIII	CLS RK	
	101.1	KVPA	III	CLS RK	
	100.3	KTEX	IIIIIIII	CTRY	
	96.1	KIWW	IIIII	SPAN	
	98.5	KGBT	IIIIIIII	SPAN	
	99.5	KKPS	IIIIIIII	SPAN	
	105.5	KTJX	III	SPAN	
	106.3	KTJN	III	SPAN	
AM	1290	KRGE	IIIII	SPAN	
AM	1530	KGBT	IIIIIIII	SPAN	
AM	1580	KIRT	I	SPAN	

	FREQ	STATION	POWER	FORMAT	
AM	1600	KBOR	I	SPAN	
	88.1	KHID	III	CLS, NPR, PRI	P
	88.9	KMBH	III	CLS, NPR, PRI	P
AM	710	KURV	IIIII	NWS, TLK	

ODESSA, TX

	FREQ	STATION	POWER	FORMAT	
	97.9	KODM	IIIII	AC	
	103.3	KCRS	IIIIIIII	AC	
	106.7	KCHX	IIIIIIII	TOP 40	
	93.3	KBAT	IIIIIIII	ROCK	
	95.1	KQRX	III	DVRS	
	96.9	KMCM	IIIII	OLDS	
	92.3	KNFM	IIIIIIII	CTRY	
	96.1	KMRK	IIIIIIII	SPAN	
	107.9	KQLM	IIIIIIII	SPAN	
AM	1070	KWEL	III	SPAN	
AM	1230	KOZA	I	SPAN	
	91.3	KOCV	III	CLS, NPR, PRI	P
AM	550	KCRS	IIIIIIII	NWS	
AM	1410	KRIL	I	NWS, TLK	

SAN ANGELO, TX

	FREQ	STATION	POWER	FORMAT	
	98.7	KELI	IIIIIIII	AC	
	94.7	KIXY	IIIII	TOP 40	
	92.9	KDCD	IIIIIIII	CTRY	
	97.5	KGKL	IIIII	CTRY	
AM	960	KGKL	IIIIIIII	CTRY	
	100.1	KYZZ	III	SPAN	
	107.5	KSJT	IIIIIIII	SPAN	
	90.1	KUTX	III	DVRS, NPR, PRI	P

SAN ANTONIO, TX

	FREQ	STATION	POWER	FORMAT	
	101.9	KQXT	IIIIIIII	AC	
	105.3	KSMG	IIIIIIII	AC	
AM	1310	KPOZ	IIIII	AC	
	96.1	KSJL	IIIII	TOP 40	
	102.7	KTFM	IIIIIIII	TOP 40	
	99.5	KISS	IIIIIIII	ROCK	
	104.5	KZEP	IIIIIIII	CLS RK	
AM	860	KONO	IIIIIIII	OLDS	
	92.1	KNBT	III	CTRY	
	97.3	KAJA	IIIIIIII	CTRY	
	100.3	KCYY	IIIIIIII	CTRY	
AM	680	KKYX	IIIIIIII	CTRY	
AM	1580	KWED	I	CTRY	
	92.9	KROM	IIIIIIII	SPAN	
	94.1	KLEY	IIIII	SPAN	
	107.5	KXTN	IIIIIIII	SPAN	
AM	720	KSAH	IIIIIIII	SPAN	
AM	1250	KZDC	I	SPAN	
AM	1350	KCOR	IIII	SPAN	
AM	1540	KEDA	IIIII	SPAN	
	106.7	KCJZ	IIIIIIII	JZZ	
AM	930	KLUP	IIIIIIII	BB, NOS	
	88.3	KPAC	IIIIIIII	CLS, NPR, PRI	P
	89.1	KSTX	IIIIIIII	NWS, TLK, NPR, PRI	P
	89.7	KWCB	III	DVRS	P
	90.1	KSYM	I	JZZ	P
	91.7	KRTU	I	CLS	C
AM	630	KSLR	IIIIIIII	RLG AC	
AM	1100	KDRY	IIIII	RLG	
AM	550	KTSA	IIIIIIII	NWS, TLK	

TX

TX

	FREQ	STATION	POWER	FORMAT
AM	760	KTKR	IIIIIIII	SPRTS
AM	1160	KENS	IIIII	NWS, TLK
AM	1200	WOAI	IIIIIIII	NWS, TLK
AM	1420	KGNB	I	NWS, TLK

TEXARKANA, TX

	FREQ	STATION	POWER	FORMAT
	107.1	KTWN	III	LGHT AC
	95.9	KPWW	III	TOP 40
	106.3	KYGL	IIIII	CLS RK
	95.1	KEWL	IIIIIIII	OLDS
	102.5	KKYR	IIIII	CTRY
AM	790	KKYR	IIIII	CTRY
AM	1400	KOWS	I	CTRY
AM	1530	KNBO	III	RLG
	103.5	KZRB	IIIII	URBAN
	104.7	KTOY	III	URBAN
AM	740	KCMC	IIIII	SPRTS

WACO, TX

	FREQ	STATION	POWER	FORMAT	
	97.5	KWTX	IIIIIIII	TOP 40	
	95.7	KCKR	IIIIIIII	CTRY	
	99.9	WACO	IIIIIIII	CTRY	
	94.5	KBCT	III	JZZ	
	107.1	KWBU	I	DVRS	C
AM	1010	KBBW	IIIII	RLG AC	
AM	1230	KWTX	I	KIDS	
AM	1460	KKTK	I	SPRTS	
AM	1580	KRZI	I	NWS, TLK	

WICHITA FALLS, TX

	FREQ	STATION	POWER	FORMAT
	102.5	KQXC	I	HOT AC
	106.3	KTLT	III	AC
	92.9	KNIN	IIIIIIII	TOP 40
	104.7	KYYI	IIIIIIII	CLS RK
AM	1290	KWFS	IIIII	OLDS
	99.9	KLUR	IIIIIIII	CTRY
	103.3	KWFS	IIIII	CTRY
AM	620	KAAM	IIIIIIII	KIDS

88.1

IA	CEDAR FALLS	KBBG	77
KS	WICHITA	KBCU	78
MN	ST. CLOUD	KVSC	82
MO	COLUMBIA	KCOU	82
MO	ST. LOUIS	KDHX	83
SD	SIOUX FALLS	KRSD	86
TX	BROWNSVILLE	KHID	91
TX	DALL.-FT. WRTH	KNTU	88
TX	LAREDO	KHOY	90
TX	LUBBOCK	KTXT	90

88.3

AR	LITTLE ROCK	KABF	76
IA	CEDAR RAPIDS	KCCK	77
LA	NEW ORLEANS	WRBH	80
TX	DALL.-FT. WRTH	KJCR	88
TX	SAN ANTONIO	KPAC	92

88.5

IA	DAVENPORT	KALA	78
IA	DES MOINES	KURE	77
TX	EL PASO	KTEP	89

88.7

LA	LAFAYETTE	KRVS	79
MN	MINN - ST. PAUL	WRFW	81
MO	JOPLIN	KXMS	82
MO	ST. LOUIS	WSIE	83
TX	AUSTIN	KAZI	87
TX	DALL.-FT. WRTH	KTCU	88
TX	HOUSTON	KUHF	89
TX	LONGVIEW	KTPB	90

88.9

MN	ST. CLOUD	KNSR	82
NE	OMAHA	KNOS	85
OK	OKLAHOMA CITY	KOCC	85
TX	BROWNSVILLE	KMBH	91

89.1

AR	LITTLE ROCK	KUAR	76
KS	WICHITA	KMUW	78
MO	ST. LOUIS	KCLC	83
SD	SIOUX FALLS	KAUR	86
TX	BRYAN	KEOS	87
TX	DALL.-FT. WRTH	KMQX	88
TX	LUBBOCK	KOHM	90
TX	SAN ANTONIO	KSTX	92

89.3

IA	DES MOINES	KUCB	77
LA	BATON ROUGE	WRKF	79
MO	KANSAS CITY	KCUR	82
ND	GRAND FORKS	KUND	84
NE	LINCOLN	KZUM	84
SD	RAPID CITY	KBHE	86
TX	DALL.-FT. WRTH	KNON	88

89.5

IA	CEDAR FALLS	KHKE	77
MO	COLUMBIA	KOPN	82
MO	ST. LOUIS	KCFV	83
OK	TULSA	KWGS	85
TX	AUSTIN	KMFA	87
TX	DALL.-FT. WRTH	KYQX	88
TX	EL PASO	KXCR	89

89.7

MO	ST. LOUIS	KYMC	82
NE	OMAHA	KIWR	85
TX	ABILENE	KACU	86
TX	HOUSTON	KACC	89
TX	SAN ANTONIO	KWCB	92

89.9

LA	NEW ORLEANS	WWNO	80
LA	SHREVEPORT	KDAQ	80
MN	DULUTH	WHSA	81
MN	MINN - ST. PAUL	KMOJ	81
MN	ROCHESTER	KRPR	81
MO	ST. LOUIS	WLCA	83
TX	AMARILLO	KACV	87

90.1

IA	DES MOINES	WOI	77
MN	ST. CLOUD	KSJR	82
OK	OKLAHOMA CITY	KCSC	85
TX	DALL.-FT. WRTH	KERA	88
TX	HOUSTON	KPFT	89
TX	SAN ANGELO	KUTX	91
TX	SAN ANTONIO	KSYM	92

90.3

IA	DAVENPORT	WVIK	78
IA	SIOUX CITY	KWIT	78
LA	MONROE	KEDM	80
MN	MINN - ST. PAUL	KFAI	81
ND	FARGO	KCCD	84
NE	LINCOLN	KRNU	84
TX	CORPUS CHRISTI	KEDT	88
TX	LONGVIEW	KBJS	90

90.5

AR	LITTLE ROCK	KLRE	76
IA	DUBUQUE	WSUP	77
MN	DULUTH	KDNI	81
MO	COLUMBIA	KWWC	82
ND	BISMARCK	KCND	84
TX	AUSTIN	KUT	87

90.7

LA	ALEXANDRIA	KLSA	79
LA	NEW ORLEANS	WWOZ	80
MN	ROCHESTER	KZSE	81
MO	ST. LOUIS	KWMU	83
ND	GRAND FORKS	KFJM	84
NE	OMAHA	KVNO	85

90.9

IA	CEDAR FALLS	KUNI	77
NE	LINCOLN	KUCV	84
SD	SIOUX FALLS	KCSD	86
TX	BRYAN	KAMU	87
TX	HOUSTON	KTSU	89

91.1

LA	BATON ROUGE	KLSU	79
LA	MONROE	KNLU	80
MN	MINN - ST. PAUL	KNOW	81
MO	SPRINGFIELD	KSMU	83
ND	FARGO	KCCM	84
TX	AMARILLO	KWTS	87

91.3

AR	FAYETTEVILLLE	KUAF	76
AR	LITTLE ROCK	KUCA	76
LA	SHREVEPORT	KSCL	80
MN	DULUTH	KUWS	81
MO	COLUMBIA	KBIA	82
OK	TULSA	KRSC	86
TX	BEAUMONT	KVLU	87
TX	HOUSTON	KPVU	89
TX	ODESSA	KOCV	91
TX	TEMPLE	KNCT	90

91.7

MN	ROCHESTER	KLSE	81
TX	AUSTIN	KVRX	87
TX	CORPUS CHRISTI	KBNJ	88
TX	HOUSTON	KTRU	89
TX	SAN ANTONIO	KRTU	92

91.9

MO	KANSAS CITY	KWJC	82
ND	FARGO	KDSU	84

92.1

AR	FAYETTEVILLLE	KKEG	76
LA	ALEXANDRIA	KLIL	78
LA	SHREVEPORT	KLKL	80
MN	DULUTH	WWAX	80
MO	COLUMBIA	KMFC	82
OK	TULSA	KOAS	85
TX	AUSTIN	KQQQ	87
TX	BRYAN	KTSR	87
TX	HOUSTON	KRTS	89
TX	LONGVIEW	KDOK	90
TX	SAN ANTONIO	KNBT	91

92.3

AR	FT. SMITH	KREU	76
AR	LITTLE ROCK	KIPR	76
KS	WICHITA	KOEZ	78
LA	NEW ORLEANS	WCKW	80
MO	KANSAS CITY	KCCV	82
MO	ST. LOUIS	WIL	83
NE	OMAHA	KEZO	84
TX	EL PASO	KOFX	89
TX	ODESSA	KNFM	91
TX	TEMPLE	KIIZ	90

92.5

AR	FT. SMITH	KPRV	76
IA	DES MOINES	KJJY	77
MN	MINN - ST. PAUL	KQRS	81
MO	JOPLIN	KSYN	82
OK	OKLAHOMA CITY	KOMA	85

SD	SIOUX FALLS	KELO	86
TX	AUSTIN	KKLB	87
TX	BEAUMONT	KTFA	87
TX	DALL.-FT. WRTH	KZPS	88

92.9

ND	FARGO	KPHT	84
TX	BROWNSVILLE	KESO	91
TX	LAREDO	KJBZ	90
TX	LUBBOCK	KJAK	90

93.1

IA	DUBUQUE	KATF	77
KS	TOPEKA	KANS	78
MN	DULUTH	WSCD	81
MN	ST. CLOUD	KKJM	82
ND	BISMARCK	KYYY	83
ND	GRAND FORKS	KKXL	84
OK	TULSA	KBEZ	85
TX	HOUSTON	KKBQ	89
TX	SAN ANGELO	KDCD	91
TX	SAN ANTONIO	KROM	91
TX	WICHITA FALLS	KNIN	92

93.3

LA	ALEXANDRIA	KQID	78
SD	RAPID CITY	KRCS	86
TX	AMARILLO	KQIZ	86
TX	EL PASO	KSII	89
TX	LONGVIEW	KTYL	90

93.5

IA	DES MOINES	KIOA	77
LA	NEW ORLEANS	WQUE	80
NE	OMAHA	KTNP	84
TX	DALL.-FT. WRTH	KKZN	88
TX	ODESSA	KBAT	91

93.7

IA	DAVENPORT	KORB	78
AR	FT. SMITH	KISR	76
LA	LAFAYETTE	KTBT	79
LA	SHREVEPORT	KITT	80
MN	MINN - ST. PAUL	KXXR	81
MO	ST. LOUIS	KSD	83
ND	FARGO	WDAY	84
TX	AUSTIN	KLBJ	87
TX	HOUSTON	KKRW	89
TX	LUBBOCK	KXTQ	90

93.9

IA	DAVENPORT	WJRE	78
KS	WICHITA	KDGS	78
LA	ALEXANDRIA	KFAD	78
MO	JOPLIN	KJMK	82
SD	RAPID CITY	KKMK	86
TX	CORPUS CHRISTI	KMXR	88
TX	EL PASO	KINT	89

94.1

AR	LITTLE ROCK	KKPT	76
MO	KANSAS CITY	KFKF	82
NE	OMAHA	WOW	85
OK	LAWTON	KZCD	85
TX	AMARILLO	KFNX	86
TX	BEAUMONT	KQXY	87
TX	DALL.-FT. WRTH	KLTY	88
TX	SAN ANTONIO	KLEY	91

94.3

AR	FAYETTEVILLLE	KAMO	76

94.5

LA	LAFAYETTE	KSMB	79
LA	SHREVEPORT	KRUF	80
MN	MINN - ST. PAUL	KSTP	81
ND	BISMARCK	KQDY	84
OK	TULSA	KEMX	86
SD	SIOUX FALLS	KCFS	86
TX	BROWNSVILLE	KFRQ	91
TX	HOUSTON	KLDE	89
TX	LUBBOCK	KFMX	90
TX	WACO	KBCT	92

94.7

LA	NEW ORLEANS	WYLA	80
MO	SPRINGFIELD	KTTS	83
MO	ST. LOUIS	KSHE	83
ND	GRAND FORKS	KNOX	84
OK	OKLAHOMA CITY	KQSR	85
TX	AUSTIN	KAMX	87
TX	CORPUS CHRISTI	KBSO	88
TX	EL PASO	KATH	89

TX	SAN ANGELO	KIXY	91

94.9

AR	FAYETTEVILLLE	KDAB	76
AR	LITTLE ROCK	KOLL	76
IA	DES MOINES	KGGO	77
LA	NEW ORLEANS	WADU	80
MN	DULUTH	KQDS	80
MN	ST. CLOUD	KMXK	82
MO	KANSAS CITY	KCMO	82
TX	DALL.-FT. WRTH	KWRD	88
TX	LAREDO	KOYE	90

95.1

KS	WICHITA	KICT	78
MO	JOPLIN	KMXL	82
NE	LINCOLN	KRKR	84
TX	BEAUMONT	KYKR	87
TX	ODESSA	KQRX	91
TX	TEXARKANA	KEWL	92

95.3

OK	LAWTON	KMGZ	85

95.5

IA	SIOUX CITY	KGLI	78
LA	LAFAYETTE	KRRQ	79
MO	SPRINGFIELD	KTOZ	83
OK	TULSA	KWEN	85
TX	AUSTIN	KKMJ	88
TX	CORPUS CHRISTI	KZFM	88
TX	EL PASO	KLAQ	89

95.7

LA	NEW ORLEANS	WTKL	80
MN	DULUTH	KDAL	80
TX	HOUSTON	KIKK	89
TX	WACO	KCKR	92

95.9

AR	FT. SMITH	KMXJ	76
TX	TEXARKANA	KPWW	92

96.1

AR	LITTLE ROCK	KSSN	76
IA	CEDAR FALLS	KCVM	77
LA	BATON ROUGE	KRVE	79
LA	LAKE CHARLES	KYKZ	79
ND	GRAND FORKS	KQHT	84
NE	OMAHA	KEFM	84
OK	OKLAHOMA CITY	KXXY	85
TX	BROWNSVILLE	KIWW	91
TX	LONGVIEW	KKTX	90
TX	ODESSA	KMRK	91
TX	SAN ANTONIO	KSJL	91

96.3

KS	WICHITA	KRZZ	78
MO	ST. LOUIS	KIHT	88
TX	DALL.-FT. WRTH	KSCS	88
TX	EL PASO	KHEY	89
TX	LUBBOCK	KLLL	90

96.5

AR	LITTLE ROCK	KHUG	76
IA	CEDAR RAPIDS	WMT	77
LA	LAFAYETTE	KFTE	79
LA	SHREVEPORT	KVKI	80
MN	ROCHESTER	KWWK	81
MO	KANSAS CITY	KXTR	82
MO	SPRINGFIELD	KLTQ	83
ND	BISMARCK	KBYZ	83
OK	TULSA	KRAV	85
TX	CORPUS CHRISTI	KLTG	88
TX	HOUSTON	KHMX	89

96.7

MN	ST. CLOUD	KKSR	81
MO	COLUMBIA	KCMQ	82
ND	FARGO	KOCL	84
TX	AUSTIN	KHFI	87

96.9

IA	DAVENPORT	WXLP	78
LA	ALEXANDRIA	KZMZ	78
TX	AMARILLO	KMML	86
TX	ODESSA	KMCM	91

97.1

LA	NEW ORLEANS	WEZB	80
MN	MINN - ST. PAUL	KTCZ	81
MO	ST. LOUIS	KXOK	83
ND	GRAND FORKS	KYCK	84
SD	RAPID CITY	KPSD	86
TX	DALL.-FT. WRTH	KEGL	88

TX	HOUSTON	KKTL	89

97.3

IA	DES MOINES	KHKI	77
IA	DUBUQUE	KGRR	77
KS	TOPEKA	WIBW	78
MN	DULUTH	KDNW	81
MO	SPRINGFIELD	KXUS	83
SD	SIOUX FALLS	KMXC	86
TX	SAN ANTONIO	KAJA	91

97.5

MN	ROCHESTER	KNXR	81
ND	BISMARCK	KKCT	84
OK	TULSA	KMOD	85
TX	BEAUMONT	KAYD	87
TX	EL PASO	KBNA	89
TX	SAN ANGELO	KGKL	91
TX	WACO	KWTX	92

97.7

IA	DUBUQUE	WGLR	77
LA	ALEXANDRIA	KAPB	79
TX	LONGVIEW	KWRW	90

97.9

AR	FT. SMITH	KZBB	76
IA	SIOUX CITY	KSEZ	78
KS	WICHITA	KRBB	78
MO	JOPLIN	KXDG	82
ND	FARGO	KFNW	84
OK	OKLAHOMA CITY	KTNT	85
SD	RAPID CITY	KLMP	86
TX	AMARILLO	KGNC	87
TX	DALL.-FT. WRTH	KBFB	88
TX	HOUSTON	KBXX	89
TX	ODESSA	KODM	91

98.1

IA	CEDAR RAPIDS	KHAK	77
LA	BATON ROUGE	WDGL	79
MN	ST. CLOUD	WWJO	82
MO	ST. LOUIS	KYKY	83
OK	LAWTON	KJMZ	85
TX	ABILENE	KKHR	86
TX	AUSTIN	KVET	87
TX	LAREDO	KRRG	90

98.3

AR	FAYETTEVILLLE	KFAY	76
LA	MONROE	KYEA	80
MO	COLUMBIA	KFMZ	82
TX	BRYAN	KORA	87
TX	CORPUS CHRISTI	KLHB	88
TX	TEMPLE	KRYL	90

98.5

AR	LITTLE ROCK	KURB	76
IA	CEDAR FALLS	KKCV	77
LA	NEW ORLEANS	WYLD	80
NE	OMAHA	KQKQ	84
OK	TULSA	KVOO	85
TX	BROWNSVILLE	KGBT	91

98.7

KS	WICHITA	KAYY	78
MO	SPRINGFIELD	KWTO	83
MO	ST. LOUIS	WJKK	83
ND	BISMARCK	KACL	83
ND	FARGO	KQWB	84
SD	RAPID CITY	KOUT	86
TX	AMARILLO	KPRF	86
TX	DALL.-FT. WRTH	KLUV	88
TX	SAN ANGELO	KELI	91

98.9

IA	DAVENPORT	WHTS	78
MN	DULUTH	KTCO	81
MN	ST. CLOUD	KZPK	82
MO	KANSAS CITY	KQRC	82
OK	OKLAHOMA CITY	KYIS	85

99.1

AR	FT. SMITH	KMAG	76
KS	WICHITA	KTLI	78
LA	LAFAYETTE	KXKC	79
MO	ST. LOUIS	KFUO	83
TX	CORPUS CHRISTI	KRYS	88
TX	DALL.-FT. WRTH	KHCK	88
TX	HOUSTON	KODA	89

99.3

IA	CEDAR FALLS	KWAY	77
IA	DUBUQUE	KDST	77

KS	TOPEKA	KWIC	78

99.5

AR	LITTLE ROCK	KYFX	76
LA	LAKE CHARLES	KHLA	79
LA	NEW ORLEANS	WRNO	80
MN	MINN - ST. PAUL	KSJN	81
MO	SPRINGFIELD	KADI	83
OK	LAWTON	KBZQ	85
TX	BROWNSVILLE	KKPS	91
TX	BRYAN	KBMA	87
TX	DALL.-FT. WRTH	KPLX	88
TX	LUBBOCK	KCRM	90
TX	SAN ANTONIO	KISS	91

99.7

LA	SHREVEPORT	KMJJ	80
MO	JOPLIN	KBTN	82
MO	KANSAS CITY	KYYS	82
TX	ABILENE	KBCY	86
TX	AMARILLO	KBZD	86

99.9

AR	FT. SMITH	KTCS	76
LA	LAFAYETTE	KTDY	79
MO	ST. LOUIS	KFAV	83
ND	FARGO	KVOX	84
NE	OMAHA	KGOR	85
TX	CORPUS CHRISTI	KSAB	88
TX	EL PASO	KTSM	89
TX	HOUSTON	KSHN	89
TX	WACO	WACO	92
TX	WICHITA FALLS	KLUR	92

100.1

TX	SAN ANGELO	KYZZ	91

100.3

AR	LITTLE ROCK	KQAR	76
IA	DES MOINES	KMXD	77
KS	TOPEKA	KDVV	78
LA	ALEXANDRIA	KRRV	79
MN	MINN - ST. PAUL	WRQC	81
MO	ST. LOUIS	KATZ	83
SD	RAPID CITY	KFXS	86
TX	BROWNSVILLE	KTEX	91
TX	DALL.-FT. WRTH	KRBV	88
TX	HOUSTON	KILT	89
TX	SAN ANTONIO	KCYY	91

100.5

OK	OKLAHOMA CITY	KATT	85
TX	LAREDO	KBDR	90

100.7

AR	FT. SMITH	KBBQ	76
LA	BATON ROUGE	WXCT	79
TX	ABILENE	KORQ	86
TX	AUSTIN	KASE	87
TX	HOUSTON	KRTX	89

100.9

LA	MONROE	KHLL	80
OK	TULSA	KXOJ	86

101.1

AR	FAYETTEVILLLE	KLRC	76
AR	LITTLE ROCK	KDRE	76
LA	NEW ORLEANS	WNOE	80
LA	SHREVEPORT	KRMD	80
MO	KANSAS CITY	KCFX	82
MO	ST. LOUIS	WVRV	83
TX	BROWNSVILLE	KVPA	91
TX	DALL.-FT. WRTH	WRR	88
TX	HOUSTON	KLOL	89
TX	LUBBOCK	KONE	90

101.3

IA	DAVENPORT	KUUL	78
KS	WICHITA	KFDI	78
LA	LAKE CHARLES	KKGB	79
MN	MINN - ST. PAUL	KDWB	81
MO	SPRINGFIELD	KTXR	83
TX	CORPUS CHRISTI	KNCN	88

101.5

LA	BATON ROUGE	WYNK	79
MO	COLUMBIA	KPLA	82
ND	BISMARCK	KSSS	83
NE	OMAHA	KISP	85
OK	LAWTON	KLAW	85
TX	LONGVIEW	KNUE	90

101.7

AR	LITTLE ROCK	KKRN	76

MN	DULUTH	KLDJ	81
MN	ROCHESTER	KRCH	81
MN	ST. CLOUD	WHMH	81
TX	TEMPLE	KLTD	90

101.9

AR	FAYETTEVILLE	KMXF	76
IA	CEDAR FALLS	KNWS	77
LA	MONROE	KNOE	80
LA	NEW ORLEANS	WLMG	80
ND	FARGO	KFGO	84
OK	OKLAHOMA CITY	KTST	85
SD	SIOUX FALLS	KTWB	86
TX	AMARILLO	KATP	87
TX	SAN ANTONIO	KQXT	91

102.1

AR	LITTLE ROCK	KOKY	76
MN	MINN - ST. PAUL	KEEY	81
MO	KANSAS CITY	KOZN	82
TX	DALL.-FT. WRTH	KTXQ	88
TX	EL PASO	KPRR	89
TX	HOUSTON	KMJQ	89

102.3

IA	DUBUQUE	KXGE	77
LA	ALEXANDRIA	KBCE	79
MO	COLUMBIA	KBXR	82
OK	TULSA	KTFX	85
TX	AUSTIN	KPEZ	87
TX	LONGVIEW	KLJT	90

102.5

AR	LITTLE ROCK	KARN	76
IA	DES MOINES	KSTZ	77
LA	BATON ROUGE	WLSS	79
MN	DULUTH	KRBR	80
MO	JOPLIN	KIXQ	82
MO	ST. LOUIS	KEZK	83
TX	BEAUMONT	KTCX	87
TX	LUBBOCK	KZII	90
TX	TEXARKANA	KKYR	92
TX	WICHITA FALLS	KQXC	92

102.7

AR	FT. SMITH	KLSZ	76
NE	LINCOLN	KFRX	84
OK	OKLAHOMA CITY	KJYO	85
SD	SIOUX FALLS	KYBB	86
TX	ABILENE	KHXS	86
TX	SAN ANTONIO	KTFM	91

102.9

IA	CEDAR RAPIDS	KZIA	77
LA	LAFAYETTE	KAJN	79
MN	MINN - ST. PAUL	WLTE	81
TX	DALL.-FT. WRTH	KDMX	88
TX	HOUSTON	KLTN	89

103.1

TX	EL PASO	KPAS	89
TX	TEMPLE	KOOV	90

103.3

IA	DUBUQUE	KIKR	77
IA	SIOUX CITY	KTFC	78
MO	KANSAS CITY	KPRS	82
MO	ST. LOUIS	KLOU	83
OK	TULSA	KJSR	85
TX	HOUSTON	KJOJ	89
TX	ODESSA	KCRS	91
TX	WICHITA FALLS	KWFS	92

103.5

LA	ALEXANDRIA	KLAA	79
MO	JOPLIN	KWXD	82
TX	AUSTIN	KEYI	87
TX	TEXARKANA	KZRB	92

103.7

AR	LITTLE ROCK	KSYG	76
IA	DAVENPORT	WLLR	78
KS	WICHITA	KEYN	78
LA	LAKE CHARLES	KBIU	79
LA	SHREVEPORT	KDKS	80
MN	ST. CLOUD	KLZZ	81
SD	SIOUX FALLS	KRRO	86
TX	ABILENE	KCDD	86
TX	CORPUS CHRISTI	KOUL	88
TX	DALL.-FT. WRTH	KVIL	88

103.9

AR	FAYETTEVILLLE	KKIX	76

104.1

IA	DES MOINES	KLTI	77
LA	MONROE	KJLO	80
MN	MINN - ST. PAUL	KMJZ	81
MO	SPRINGFIELD	KZRQ	83
MO	ST. LOUIS	WXTM	83
OK	OKLAHOMA CITY	KMGL	85
SD	RAPID CITY	KIQK	86
TX	BROWNSVILLE	KBFM	90
TX	HOUSTON	KRBE	89
TX	LONGVIEW	KKUS	90

104.3

LA	ALEXANDRIA	KEZP	79
MO	KANSAS CITY	KBEQ	82
ND	GRAND FORKS	KZLT	84
TX	AUSTIN	KQBT	87

104.5

IA	CEDAR RAPIDS	KDAT	77
KS	WICHITA	KLLS	78
NE	OMAHA	KSRZ	84
OK	TULSA	KMYZ	85
TX	BEAUMONT	KKMY	87
TX	DALL.-FT. WRTH	KKDA	88
TX	SAN ANTONIO	KZEP	91

104.7

LA	LAFAYETTE	KNEK	79
LA	NEW ORLEANS	WYLK	80
MN	ST. CLOUD	KCLD	81
MO	SPRINGFIELD	KKLH	83
ND	BISMARCK	KNDR	84
SD	SIOUX FALLS	KKLS	86
TX	BRYAN	KKYS	87
TX	TEXARKANA	KTOY	92
TX	WICHITA FALLS	KYYI	92

104.9

AR	FAYETTEVILLLE	KBRS	76
LA	BATON ROUGE	KKAY	79
MO	ST. LOUIS	KMJM	83
OK	OKLAHOMA CITY	KNTL	85
OK	TULSA	KREK	85
TX	HOUSTON	KOVA	89

105.1

AR	LITTLE ROCK	KMJX	76
IA	DES MOINES	KCCQ	77
MN	DULUTH	KKCB	81
MN	MINN - ST. PAUL	KZNR	81
MO	SPRINGFIELD	KOSP	83
TX	ABILENE	KEAN	86
TX	CORPUS CHRISTI	KMIQ	88

105.3

IA	DUBUQUE	KLYV	77
KS	WICHITA	KWSJ	78
LA	LAKE CHARLES	KZWA	79
LA	MONROE	KLIP	80
LA	NEW ORLEANS	WLTS	80
MN	MINN - ST. PAUL	KZNT	81
MN	ROCHESTER	KYBA	81
NE	LINCOLN	KKUL	84
OK	TULSA	KJMM	86
TX	DALL.-FT. WRTH	KYNG	88
TX	SAN ANTONIO	KSMG	91

105.5

LA	LAFAYETTE	KJJB	79
MN	ST. CLOUD	KDDG	82
TX	BROWNSVILLE	KTJX	91
TX	CORPUS CHRISTI	KRAD	88
TX	TEMPLE	KYUL	90

105.7

AR	FAYETTEVILLLE	KMCK	76
IA	CEDAR FALLS	KOKZ	77
MN	MINN - ST. PAUL	KZNZ	81
OK	OKLAHOMA CITY	KROU	85
TX	AMARILLO	KAEZ	86
TX	LONGVIEW	KYKX	90
TX	LUBBOCK	KRBL	90

105.9

LA	LAFAYETTE	KVOL	79
MO	SPRINGFIELD	KGBX	83
NE	OMAHA	KKCD	84

106.1

LA	MONROE	KMYY	80
MO	COLUMBIA	KOQL	82
OK	TULSA	KQLL	85
TX	BEAUMONT	KIOC	87

D

TX	DALL.-FT. WRTH	KHKS	88
TX	LAREDO	KNEX	90

106.3

AR	FT. SMITH	KZKZ	76
AR	LITTLE ROCK	KHTE	76
IA	DES MOINES	KYSY	77
NE	LINCOLN	KIBZ	84
OK	OKLAHOMA CITY	KGOU	85
TX	ABILENE	KFQX	86
TX	BROWNSVILLE	KTJN	91
TX	TEMPLE	KOOC	90
TX	TEXARKANA	KYGL	92
TX	WICHITA FALLS	KTLT	92

106.5

AR	FAYETTEVILLLE	KBVA	76
IA	DAVENPORT	KCQQ	78
MO	KANSAS CITY	KCIY	82
MO	ST. LOUIS	WKKX	83
TX	HOUSTON	KQQK	89
TX	LONGVIEW	KOOI	90
TX	LUBBOCK	KEJS	90

106.7

AR	LITTLE ROCK	KDDK	76
LA	LAFAYETTE	KLTW	79
MO	ST. LOUIS	WSTZ	83
TX	DALL.-FT. WRTH	KDXT	88
TX	ODESSA	KCHX	91
TX	SAN ANTONIO	KCJZ	92

106.9

KS	TOPEKA	KTPK	78
MN	ROCHESTER	KROC	81
TX	DALL.-FT. WRTH	KZDF	88
TX	HOUSTON	KKHT	89

107.1

IA	DUBUQUE	WPVL	77
IA	SIOUX CITY	KSFT	78
LA	LAFAYETTE	KOGM	79
MN	MINN - ST. PAUL	WIXK	81
ND	GRAND FORKS	KKEQ	84
TX	AMARILLO	KPUR	86
TX	AUSTIN	KGSR	87
TX	DALL.-FT. WRTH	KZDL	88
TX	TEXARKANA	KTWN	92
TX	WACO	KWBU	92

107.3

KS	WICHITA	KKRD	74
LA	BATON ROUGE	WTGE	79
MO	KANSAS CITY	KNRX	82
NE	LINCOLN	KEZG	84
OK	LAWTON	KVRW	85
TX	CORPUS CHRISTI	KCCG	88
TX	LONGVIEW	KISX	90
TX	TEMPLE	KLFX	90

107.5

IA	DES MOINES	KKDM	77
IA	DUBUQUE	WJOD	77
ND	GRAND FORKS	KJKJ	84
TX	DALL.-FT. WRTH	KOAI	88
TX	HOUSTON	KTBZ	89
TX	SAN ANGELO	KSJT	91
TX	SAN ANTONIO	KXTN	91

107.7

AR	LITTLE ROCK	KLAL	76
KS	TOPEKA	KMAJ	78
MN	DULUTH	KUSZ	81
MO	ST. LOUIS	KSLZ	83
OK	OKLAHOMA CITY	KRXO	85
TX	AUSTIN	KAHK	87

107.9

AR	FAYETTEVILLLE	KEZA	76
IA	CEDAR FALLS	KFMW	77
MN	MINN - ST. PAUL	KQQL	81
ND	FARGO	KPFX	84
TX	ABILENE	KEYJ	86
TX	AMARILLO	KZRK	86
TX	BROWNSVILLE	KVLY	90
TX	ODESSA	KQLM	91

AM

540

LA	MONROE	KNOE	80

550

MO	ST. LOUIS	KTRS	83

ND	BISMARCK	KFYR	83
TX	ODESSA	KCRS	91
TX	SAN ANTONIO	KTSA	92

560

MN	DULUTH	WEBC	81
MO	SPRINGFIELD	KWTO	83
TX	BEAUMONT	KLVI	87

570

TX	DALLAS-FT. WORTH	KLIF	88

580

KS	TOPEKA	WIBW	78
LA	ALEXANDRIA	KLBG	79
TX	LUBBOCK	KRFE	90

590

MO	ST. LOUIS	KFNS	83
NE	OMAHA	WOW	85
TX	AUSTIN	KLBJ	87

600

IA	CEDAR RAPIDS	WMT	77
TX	EL PASO	KROD	89
TX	LONGVIEW	KTBB	90

610

MN	DULUTH	KDAL	80
MO	KANSAS CITY	WDAF	82
TX	HOUSTON	KILT	89

620

IA	SIOUX CITY	KMNS	78
TX	WICHITA FALLS	KAAM	92

630

MO	ST. LOUIS	KJSL	83
TX	SAN ANTONIO	KSLR	92

640

IA	DES MOINES	WOI	77
OK	OKLAHOMA CITY	WWLS	85

660

NE	OMAHA	KCRO	85

680

TX	SAN ANTONIO	KKYX	91

690

LA	NEW ORLEANS	WTIX	80
TX	EL PASO	KHEY	89
TX	LONGVIEW	KZEY	90

700

TX	HOUSTON	KSEV	89

710

MN	DULUTH	WDSM	81
MO	KANSAS CITY	KCMO	82
TX	AMARILLO	KGNC	87
TX	BROWNSVILLE	KURV	91

720

TX	SAN ANTONIO	KSAH	

730

MO	ST LOUIS		
TX	DALLAS-FT. WORTH		

740

OK	TULSA	KRMG	86
TX	HOUSTON	KTRH	89
TX	TEXARKANA	KCMC	92

750

TX	EL PASO	KAMA	89

760

TX	SAN ANTONIO	KTKR	92

770

LA	LAFAYETTE	KJCB	79
MN	MINNEAPOLIS-ST. P	KUOM	81

790

AR	FAYETTEVILLLE	KURM	76
ND	FARGO	KFGO	84
TX	HOUSTON	KBME	89
TX	LUBBOCK	KFYO	90
TX	TEXARKANA	KKYR	92

800

LA	NEW ORLEANS	WSHO	80
OK	OKLAHOMA CITY	KQCV	85

810

MO	KANSAS CITY	WHB	82

820

TX	DALLAS-FT. WORTH	WBAP	88

830

LA	NEW ORLEANS	WFNO	80
MN	MINNEAPOLIS-ST. P	WCCO	81

860

TX	SAN ANTONIO	KONO	91

870

LA	NEW ORLEANS	WWL	80

900

MN	MINNEAPOLIS-ST. P	KTIS	81

910

LA	BATON ROUGE	WNDC	79

920

AR	LITTLE ROCK	KARN	76
MO	ST. LOUIS	WGNU	83
SD	RAPID CITY	KKLS	86
TX	EL PASO	KBNA	89

930

OK	OKLAHOMA CITY	WKY	85
TX	SAN ANTONIO	KLUP	92

940

IA	DES MOINES	KXTK	77
LA	NEW ORLEANS	WYLD	80
TX	AMARILLO	KIXZ	87

950

AR	FT. SMITH	KFSA	76
KS	WICHITA	KJRG	78
TX	HOUSTON	KPRC	89
TX	LUBBOCK	KXTQ	90

960

TX	SAN ANGELO	KGKL	91

970

LA	ALEXANDRIA	KSYL	79
ND	FARGO	WDAY	84
OK	TULSA	KCFO	86
TX	AUSTIN	KIXL	87
TX	DALLAS-FT. WORTH	KHVN	88

980

MN	MINNEAPOLIS-ST. P	KKMS	81
MO	KANSAS CITY	KMBZ	82

990

LA	NEW ORLEANS	WGSO	80

1000

OK	OKLAHOMA CITY	KTOK	85

1010

TX	HOUSTON	KLAT	89
TX	WACO	KBBW	92

1030

AR	FAYETTEVILLLE	KFAY	76
MO	KANSAS CITY	KOWW	82

1040

IA	DES MOINES	WHO	77

1060

LA	NEW ORLEANS	WLNO	80
TX	EL PASO	KFNA	89

1070

KS	WICHITA	KFDI	78
TX	ODESSA	KWEL	91

1080

TX	DALLAS-FT. WORTH	KRLD	88

1090

AR	LITTLE ROCK	KAAY	76
IA	CEDAR FALLS	KNWS	77

1100

TX	SAN ANTONIO	KDRY	92

1110

NE	OMAHA	KFAB	85

1120

MO	ST. LOUIS	KMOX	83

1130

LA	SHREVEPORT	KWKH	80
MN	MINNEAPOLIS-ST. P	KFAN	81
ND	BISMARCK	KBMR	84

1140

SD	SIOUX FALLS	KSOO	86

1150

AR	LITTLE ROCK	KLRG	76
IA	DES MOINES	KWKY	77
LA	BATON ROUGE	WJBO	79
SD	RAPID CITY	KIMM	86
TX	BRYAN	WTAW	87
TX	CORPUS CHRISTI	KCCT	88
TX	EL PASO	KSVE	89

1160

TX	SAN ANTONIO	KENS	92

1170

IA	DAVENPORT	KJOC	78
OK	TULSA	KVOO	85

1180

1190

NE	OMAHA	KOIL	85
MO	KANSAS CITY	KPHN	82
TX	DALLAS-FT. WORTH	KLUV	88

1200

ND	FARGO	KFNW	84
TX	SAN ANTONIO	WOAI	92

1210

LA	BATON ROUGE	WSKR	79

1220

MN	MINNEAPOLIS-ST. P	WEZU	81
MO	ST. LOUIS	KLPW	83

1230

AR	FT. SMITH	KFPW	76
IA	DAVENPORT	WLLR	78
LA	LAFAYETTE	KSLO	79
LA	MONROE	KLIC	80
LA	NEW ORLEANS	WBOK	80
MO	JOPLIN	KWAS	82
SD	SIOUX FALLS	KWSN	86
TX	CORPUS CHRISTI	KSIX	88
TX	ODESSA	KOZA	91
TX	WACO	KWTX	92

1240

KS	WICHITA	KNSS	78
LA	LAFAYETTE	KANE	79
LA	SHREVEPORT	KASO	80
MN	ST. CLOUD	WJON	82
NE	LINCOLN	KFOR	84
TX	BRYAN	KTAM	87

1250

AR	LITTLE ROCK	KLIH	76
IA	CEDAR RAPIDS	KCNZ	77
TX	BEAUMONT	KALO	87
TX	SAN ANTONIO	KZDC	91

1260

LA	BATON ROUGE	KBRH	79
MO	SPRINGFIELD	KTTS	83
MO	ST. LOUIS	WIBV	83
ND	GRAND FORKS	KROX	84
TX	AUSTIN	KTAE	87

1270

IA	DAVENPORT	WKBF	78
MN	ROCHESTER	KWEB	81
ND	BISMARCK	KLXX	84
TX	DALLAS-FT. WORTH	KESS	88

1280

AR	FT. SMITH	KPRV	76
LA	NEW ORLEANS	WODT	80
MN	MINNEAPOLIS-ST. P	WWTC	81
ND	FARGO	KVOX	84
TX	ABILENE	KGMM	86

1290

AR	FAYETTEVILLLE	KUOA	76
NE	OMAHA	KKAR	85
TX	BROWNSVILLE	KRGE	91
TX	WICHITA FALLS	KWFS	92

1300

LA	BATON ROUGE	WIBR	79
OK	TULSA	KAKC	86
TX	AUSTIN	KVET	87
TX	LAREDO	KLAR	90

1310

IA	DES MOINES	KDLS	77
ND	GRAND FORKS	KNOX	84
TX	AMARILLO	KZIP	87
TX	DALLAS-FT. WORTH	KTCK	88
TX	SAN ANTONIO	KPOZ	91

1320

AR	FT. SMITH	KWHN	76
LA	SHREVEPORT	KNCB	80
MO	ST. LOUIS	KSIV	83
SD	SIOUX FALLS	KELO	86
TX	HOUSTON	KXYZ	89

1330

IA	CEDAR FALLS	KWLO	77
KS	WICHITA	KFH	78
LA	LAFAYETTE	KVOL	79
MN	MINNEAPOLIS-ST. P	WMNN	81

1340

LA	SHREVEPORT	KRMD	80
MN	ROCHESTER	KROC	81
MO	KANSAS CITY	KFEZ	82

OK	OKLAHOMA CITY	KEBC	85
OK	TULSA	KTOW	86
SD	RAPID CITY	KTOQ	86
TX	ABILENE	KYYD	86
TX	BEAUMONT	KOLE	87
TX	EL PASO	KVIV	89
TX	HOUSTON	KKAM	89

1350

IA	DES MOINES	KRNT	77
LA	NEW ORLEANS	WSMB	80
TX	SAN ANTONIO	KCOR	92

1360

IA	CEDAR RAPIDS	KTOF	77
IA	SIOUX CITY	KSCJ	78
LA	LAFAYETTE	KNIR	79
TX	AMARILLO	KDJW	87
TX	CORPUS CHRISTI	KRYS	88
TX	HOUSTON	KWWJ	89

1370

IA	DUBUQUE	KDTH	77
ND	GRAND FORKS	KUND	84
TX	AUSTIN	KJCE	87
TX	LONGVIEW	KFRO	90

1380

OK	LAWTON	KXCA	85
SD	RAPID CITY	KOTA	86
TX	EL PASO	KTSM	89

1390

MN	ST. CLOUD	KXSS	82

1400

LA	LAKE CHARLES	KAOK	79
MN	MINNEAPOLIS-ST. P	KLBB	81
MO	COLUMBIA	KFRU	82
MO	SPRINGFIELD	KGMY	83
NE	LINCOLN	KLIN	84
OK	OKLAHOMA CITY	KNOR	85
TX	CORPUS CHRISTI	KUNO	88
TX	LONGVIEW	KEBE	90
TX	TEMPLE	KTEM	90
TX	TEXARKANA	KOWS	92

1410

AR	FT. SMITH	KTCS	76
TX	ODESSA	KRIL	91

1420

IA	DAVENPORT	WOC	78
LA	LAFAYETTE	KPEL	79
MO	JOPLIN	KBTN	82
NE	OMAHA	KBBX	85
TX	LUBBOCK	KLFB	90
TX	SAN ANTONIO	KGNB	92

1430

MO	ST. LOUIS	WRTH	83
OK	TULSA	KQLL	86
TX	LONGVIEW	KEES	90

1440

AR	LITTLE ROCK	KITA	76
KS	TOPEKA	KMAJ	78
LA	MONROE	KMLB	80
ND	GRAND FORKS	KKXL	84
TX	AMARILLO	KPUR	87
TX	AUSTIN	KELG	87
TX	CORPUS CHRISTI	KEYS	88

1450

IA	CEDAR RAPIDS	KMRY	77
IA	DAVENPORT	WKEI	78
LA	LAFAYETTE	KSIG	79
LA	NEW ORLEANS	WBYU	80
MN	ST. CLOUD	KNSI	82
MO	JOPLIN	WMBH	82
OK	OKLAHOMA CITY	KGFF	85
TX	BEAUMONT	KAYD	87

1460

LA	BATON ROUGE	WXOK	79
MO	ST. LOUIS	KIRL	83
TX	LUBBOCK	KBZO	90
TX	WACO	KKTK	92

1470

IA	SIOUX CITY	KWSL	78
LA	LAKE CHARLES	KLCL	79
TX	ABILENE	KBBA	86

1480

KS	WICHITA	KQAM	78
ND	GRAND FORKS	KKCQ	84

NE	LINCOLN	KLMS	84
TX	DALLAS-FT. WORTH	KDXX	88

1490

IA	DUBUQUE	WDBQ	77
KS	TOPEKA	KTOP	78
LA	LAFAYETTE	KEUN	79
MO	JOPLIN	KDMO	82
MO	ST. LOUIS	WESL	83
NE	OMAHA	KOSR	85
TX	AUSTIN	KFON	87
TX	LAREDO	KLNT	90
TX	LONGVIEW	KYZS	90

1500

MN	MINNEAPOLIS-ST. P	KSTP	81

1520

LA	LAFAYETTE	KDYS	79
OK	OKLAHOMA CITY	KOMA	85

1530

TX	BROWNSVILLE	KGBT	91
TX	TEXARKANA	KNBO	92

1540

IA	CEDAR FALLS	KXEL	77
TX	SAN ANTONIO	KEDA	92

1550

ND	FARGO	KQWB	84

1560

MO	JOPLIN	KQYX	82
TX	AUSTIN	KTXZ	87

1580

LA	LAKE CHARLES	KXZZ	79
TX	BROWNSVILLE	KIRT	91
TX	SAN ANTONIO	KWED	91
TX	WACO	KRZI	92

1590

IA	DUBUQUE	WPVL	77
MN	MINNEAPOLIS-ST. P	WIXK	81
ND	GRAND FORKS	KCNN	84
TX	CORPUS CHRISTI	KDAE	88
TX	EL PASO	KELP	89
TX	HOUSTON	KYOK	89
TX	LUBBOCK	KDAV	90

1600

IA	CEDAR RAPIDS	KCRG	77
MO	ST. LOUIS	KATZ	83
TX	BEAUMONT	KOGT	87
TX	BROWNSVILLE	KBOR	91
TX	DALLAS-FT. WORTH	KRVA	88

D

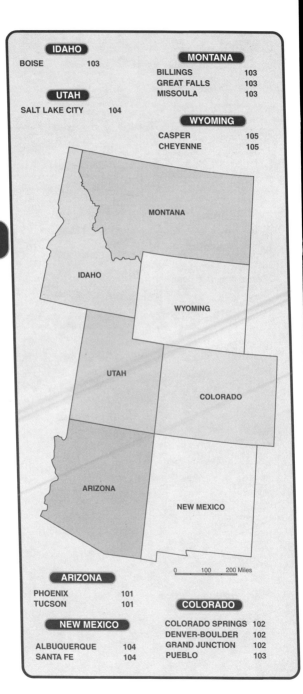

IDAHO

BOISE 103

UTAH

SALT LAKE CITY 104

MONTANA

BILLINGS 103
GREAT FALLS 103
MISSOULA 103

WYOMING

CASPER 105
CHEYENNE 105

ARIZONA

PHOENIX 101
TUCSON 101

NEW MEXICO

ALBUQUERQUE 104
SANTA FE 104

COLORADO

COLORADO SPRINGS 102
DENVER-BOULDER 102
GRAND JUNCTION 102
PUEBLO 103

E

	FREQ	STATION	POWER	FORMAT	
	96.9	KMXP	IIIIIII	HOT AC	
	98.7	KKLT	IIIIIII	AC	
	99.9	KESZ	IIIIIII	AC	
	92.3	KKFR	IIIIIII	TOP 40	
	104.7	KZZP	IIIIIII	TOP 40	
	93.3	KDKB	IIIIIII	ROCK	
	97.9	KUPD	IIIIIII	MOD RK	
	100.7	KSLX	IIIIIII	CLS RK	
	101.5	KZON	IIIIIII	DVRS	
	106.3	KEDJ	III	DVRS	
AM	1440	KSLX	IIIII	CLS RK	
	94.5	KOOL	IIIIIII	OLDS	
	102.5	KNIX	IIIIIII	CTRY	
	103.5	KWCY	IIIIIII	CTRY	
	107.9	KMLE	IIIIIII	CTRY	
AM	1230	KISO	I	CTRY	
AM	1400	KSUN	I	SPAN	
AM	1480	KPHX	IIIII	SPAN	
	95.5	KYOT	IIIII	JZZ	
	105.3	KMYL	IIIIIII	BB, NOS	
AM	550	KOY	IIIIIII	BB, NOS	
AM	1190	KMYL	IIIII	BB, NOS	
	89.5	KBAQ	III	CLS, NPR, PRI	C
	91.5	KJZZ	IIIIIII	JZZ, NPR, PRI	P
AM	960	KPXQ	IIIIIII	RLG	
AM	1580	KCWW	IIIIIII	KIDS	
	106.9	KMJK	III	URBAN	
AM	620	KTAR	IIIIIII	NWS, TLK	
AM	860	KMVP	III	SPRTS	
AM	910	KFYI	IIIIIII	NWS, TLK	
AM	1060	KDUS	IIIII	SPRTS	
AM	1310	KXAM	IIIII	NWS, TLK	
AM	1360	KGME	IIIII	SPRTS	
AM	1510	KFNN	IIIII	BUS NWS	

TUCSON, AZ

	FREQ	STATION	POWER	FORMAT	
	94.9	KMXZ	IIIIIII	AC	
	104.1	KZPT	I	AC	
	93.7	KRQQ	IIIIIII	TOP 40	
	98.3	KOHT	I	TOP 40	
	92.1	KFMA	IIIIIII	DVRS	
	96.1	KLPX	IIIIIII	ROCK	
	107.5	KHYT	IIIIIII	CLS RK	
	92.9	KWFM	IIIIIII	OLDS	
	99.5	KIIM	IIIIIII	CTRY	
AM	1290	KCUB	I	CTRY	
AM	1030	KEVT	IIIII	SPAN	
AM	1210	KQTL	IIIII	SPAN	
AM	1450	KTZR	I	SPAN	
AM	1600	KXEW	I	SPAN	
	97.5	KOAZ	III	JZZ	
AM	580	KSAZ	IIIIIII	BB, NOS	
AM	940	KCEE	III	BB, NOS	
	89.1	KUAZ	I	JZZ, NPR, PRI	P
	90.5	KUAT	IIIII	CLS, NPR, PRI	P
	91.3	KXCI	III	DVRS	P
AM	1550	KUAT	IIIIIII	JZZ, NPR, PRI	P
	97.1	KGMS	III	RLG AC	
AM	830	KFLT	IIIIIII	RLG	P
AM	1400	KTUC	I	EZ	
AM	790	KNST	IIIIIII	NWS, TLK	
AM	990	KTKT	IIIIIII	NWS, TLK	

	FREQ	STATION	POWER	FORMAT	
AM	1490	KFFN	I	SPRTS	

COLORADO SPRINGS, CO

	FREQ	STATION	POWER	FORMAT	
	95.1	KRDO	IIIIIIII	LGHT AC	
	106.3	KKLI	IIIIIIII	AC	
	94.3	KILO	IIIIIIII	ROCK	
	98.1	KKFM	IIIIIIII	CLS RK	
	92.9	KSPZ	IIIIIIII	OLDS	
	101.9	KKCS	IIIIIIII	CTRY	
	105.5	KSKX	III	JZZ	
AM	740	KTWK	IIIIIIII	BB, NOS	
	88.7	KCME	IIIII	CLS	P
	89.7	KEPC	III	DVRS	C
	91.5	KRCC	III	DVRS, NPR, PRI	C
AM	1040	KCBR	IIIII	RLG AC	
	96.1	KPRZ	III	RLG	
	102.7	KBIQ	IIIIIIII	RLG	
AM	1240	KRDO	I	SPRTS	
AM	1300	KVOR	IIIII	NWS, TLK	
AM	1460	KKCS	IIIII	NWS, TLK	

DENVER, CO

	FREQ	STATION	POWER	FORMAT	
	100.3	KIMN	IIIIIIII	AC	
	101.1	KOSI	IIIIIIII	AC	
	105.9	KALC	IIIIIIII	HOT AC	
	96.5	KXPK	III	DVRS	
	97.3	KBCO	IIIIIIII	ALT RK	
	99.5	KKHK	IIIII	CLS RK	
	103.5	KRFX	IIIII	CLS RK	
	106.7	KBPI	IIIIIIII	MOD RK	
	105.1	KXKL	IIIIIIII	OLDS	
	98.5	KYGO	IIIIIIII	CTRY	
	104.3	KCKK	IIIIIIII	CTRY	
	92.1	KJMN	IIIIIIII	SPAN	
AM	1090	KMXA	IIIIIIII	SPAN	
AM	1150	KCUV	IIIII	SPAN	
AM	1390	KJME	IIIII	SPAN	
	95.7	KHIH	IIIIIIII	JZZ	
AM	1430	KEZW	IIIII	BB, NOS	
	88.5	KGNU	I	ECLC, NPR, PRI	P
	89.3	KUVO	IIIII	JZZ, NPR, PRI	P
	90.1	KCFR	IIIIIIII	CLS, NPR, PRI	P
	94.7	KRKS	IIIIIIII	RLG AC	
AM	560	KLZ	IIIIIIII	RLG AC	
AM	990	KRKS	IIIIIIII	RLG	
	107.5	KQKS	III	URBAN	
AM	1510	KDKO	IIIII	URBAN	
AM	630	KHOW	IIIIIIII	NWS, TLK	
AM	710	KNUS	IIIIIIII	NWS	
AM	760	KTLK	IIIIIIII	NWS, TLK	
AM	850	KOA	IIIIIIII	NWS, TLK	
AM	950	KKFN	IIIIIIII	SPRTS	
AM	1280	KRRF	IIIII	NWS, TLK	

GRAND JUNCTION, CO

	FREQ	STATION	POWER	FORMAT	
	92.3	KJYE	IIIIIIII	LGHT AC	
	93.1	KQIX	IIIIIIII	HOT AC	
	104.3	KMXY	IIIIIIII	HOT AC	
	107.9	KBKL	IIIIIIII	OLDS	
	99.9	KEKB	IIIIIIII	CTRY	
	89.5	KPRN	IIIII	CLS, NPR, PRI	P
	91.3	KMSA	I	DVRS	C
AM	1230	KEXO	I	RLG AC	
AM	1100	KNZZ	IIIIIIII	NWS, TLK	

CO

	FREQ	STATION	POWER	FORMAT	
AM	1340	KQIL	I	SPRTS	
		PUEBLO			
	104.5	KYZX	III	CLS RK	
	107.9	KDZA	III	OLDS	
	96.9	KCCY	IIIII	CTRY	
	107.1	KNKN	IIIII	SPAN	
AM	1350	KGHF	I	BB, NOS	
	89.5	KTSC	I	JZZ	C
	91.9	KCFP	III	CLS, NPR, PRI	P
AM	970	KFEL	IIIII	RLG	
AM	590	KCSJ	IIIIIIII	NWS, TLK	
AM	1230	KKPC	I	NWS	
		BOISE, ID			
	105.9	KCIX	IIIIIIII	HOT AC	
	107.9	KXLT	IIIIIIII	LGHT AC	
	94.9	KFXJ	IIIIIIII	ALT RK	
	96.9	KKGL	IIIIIIII	CLS RK	
	103.3	KARO	IIIIIIII	CLS RK	
	105.1	KJOT	IIIIIIII	CLS RK	
	107.1	KCID	IIIIIIII	MOD RK	
	104.3	KLTB	IIIIIIII	OLDS	
AM	1140	KGEM	IIIII	OLDS	
	92.3	KIZN	IIIIIIII	CTRY	
	97.9	KQFC	IIIIIIII	CTRY	
AM	580	KFXD	IIIIIIII	CTRY	
	90.3	KBSU	IIIII	CLS, NPR, PRI	P
	91.5	KBSX	III	NWS, TLK, NPR, PRI	P
	730	KBSU	IIIIIIII	JZZ, NPR, PRI	P
	94.1	KBXL	IIIIIIII	RLG AC	
AM	630	KIDO	IIIIIIII	NWS, TLK	
AM	670	KBOI	IIIIIIII	NWS, TLK	
AM	1340	KTIK	I	SPRTS	
		BILLINGS, MT			
	93.3	KYYA	IIIIIIII	HOT AC	
	103.7	KBBB	IIIII	AC	
	101.7	KRSQ	IIIII	TOP 40	
	94.1	KRKX	IIIIIIII	CLS RK	
	97.1	KKBR	III	OLDS	
	98.5	KIDX	IIIII	CTRY	
	102.9	KCTR	IIIII	CTRY	
AM	790	KGHL	IIIIIIII	CTRY	
AM	970	KBUL	IIIIIIII	CTRY	
	91.7	KEMC	IIIIIIII	CLS, NPR, PRI	P
AM	1240	KMZK	I	RLG AC	
AM	730	KURL	IIIIIIII	RLG	
AM	1490	KBSR	I	NWS, TLK	
		GREAT FALLS, MT			
	98.9	KAAK	IIIII	AC	
	106.1	KQDI	IIIII	CLS RK	
	92.9	KLFM	IIIII	OLDS	
	94.5	KMON	IIIII	CTRY	
AM	560	KMON	IIIIIIII	CTRY	
AM	1310	KEIN	IIIII	CTRY	
AM	1400	KXGF	I	BB, NOS	
	89.9	KGPR	III	CLS, NPR, PRI	P
AM	1450	KQDI	I	NWS, TLK	
		MISSOULA, MT			
	102.5	KMSO	IIIII	AC	
	100.1	KZOQ	III	CLS RK	

CO
ID
MT

	FREQ	STATION	POWER	FORMAT	
AM	1340	KYLT	I	OLDS	
	93.3	KGGL	IIIIIII	CTRY	
	94.9	KYSS	IIIIIII	CTRY	
	89.1	KUFM	IIIII	CLS, NPR, PRI	P
	89.9	KBGA	III	DVRS	C
AM	930	KLCY	IIIIIII	NWS, TLK	
AM	1290	KGVO	IIIII	NWS, TLK	
AM	1450	KGRZ	I	SPRTS	

ALBUQUERQUE, NM

	FREQ	STATION	POWER	FORMAT	
	93.3	KKOB	IIIII	AC	
	99.5	KMGA	IIIII	LGHT AC	
	100.3	KPEK	IIIIIII	AC	
	106.3	KYLZ	IIIIIII	AC	
AM	920	KHTL	III	HOT AC	
	94.1	KZRR	IIIII	ROCK	
	101.3	KEZF	III	ROCK	
	102.5	KIOT	IIIII	CLS RK	
	107.9	KTEG	IIIIIII	DVRS	
	101.7	KZKL	III	OLDS	
	92.3	KRST	IIIII	CTRY	
	103.3	KTBL	IIIIIII	CTRY	
	97.7	KLVO	IIIII	SPAN	
	104.7	KEXT	III	SPAN	
AM	860	KARS	III	SPAN	
AM	1190	KXKS	IIIII	SPAN	
AM	1240	KALY	I	SPAN	
AM	1350	KABQ	IIIII	SPAN	
AM	1450	KRZY	I	SPAN	
	96.3	KHFM	IIIII	CLS	
AM	1310	KIVA	IIIII	BB, NOS	
	89.9	KUNM	IIIII	NWS, TLK, NPR	P
	107.1	KNKT	IIIII	RLG AC	
AM	1000	KKIM	IIIIIII	RLG AC	
AM	610	KZSS	IIIIIII	NWS, TLK	
AM	770	KKOB	IIIIIII	NWS, TLK	
AM	1050	KNML	I	SPRTS	
AM	1150	KDEF	IIIII	SPRTS	
AM	1240	KFBC	I	NWS, TLK	

SANTA FE, NM

	FREQ	STATION	POWER	FORMAT	
	95.5	KMMG	IIIIIII	AC	
	101.1	KSFQ	I	70's RK	
	98.5	KABG	IIIIIII	OLDS	
	106.7	KBOM	IIIII	OLDS	
AM	1400	KTRC	I	BB, NOS	
	90.7	KSFR	I	CLS	C
AM	1260	KVSF	IIIII	NWS, TLK	
AM	1490	KRSN	I	NWS, TLK	

SALT LAKE CITY, UT

	FREQ	STATION	POWER	FORMAT	
	97.1	KISN	IIIIIII	AC	
	98.7	KBEE	IIIIIII	AC	
	100.3	KSFI	IIIIIII	AC	
	102.7	KQMB	IIIIIII	HOT AC	
	105.7	KUMT	IIIIIII	LGHT AC	
	106.5	KOSY	III	LGHT AC	
	107.5	KENZ	IIIIIII	AC	
	107.9	KSNU	IIIIIII	AC	
	94.9	KZHT	IIIIIII	TOP 40	
	96.3	KXRK	IIIIIII	DVRS	
	99.5	KURR	IIIIIII	ROCK	
	101.1	KBER	IIIIIII	ROCK	

MT
NM
UT

	FREQ	STATION	POWER	FORMAT	
	103.5	KRSP	‖‖‖‖‖‖	CLS RK	
	94.1	KODJ	‖‖‖‖‖‖	OLDS	
	93.3	KUBL	‖‖‖‖‖‖	CTRY	
	101.9	KKAT	‖‖	CTRY	
	101.9	KMUS	‖‖‖	CTRY	
	104.3	KSOP	‖‖‖‖‖‖	CTRY	
AM	1370	KSOP	‖‖‖	CTRY	
	97.9	KBZN	‖‖‖‖‖‖	JZZ	
AM	1370	KJJL	‖	BB, NOS	
AM	960	KOVO	‖‖‖‖‖‖	BB, NOS	
AM	1060	KKDS	‖‖‖	BB, NOS	
AM	1280	KDYL	‖‖‖	BB, NOS	
AM	1430	KLO	‖‖‖	BB, NOS	
	88.1	KWCR	‖	DVRS	C
	88.3	KCPW	‖‖	NWS, TLK, NPR, PRI	P
	89.1	KBYU	‖‖‖‖‖‖	CLS, NPR, PRI	P
	90.1	KUER	‖‖‖‖‖‖	CLS, NPR, PRI	P
	90.9	KRCL	‖‖‖	DVRS	P
AM	860	KCNR	‖‖‖‖‖‖	KIDS	
AM	1400	KSRR	‖	FS	
AM	570	KNRS	‖‖‖‖‖‖	SPRTS	
AM	630	KTKK	‖‖‖	NWS, TLK	
AM	650	KGAB	‖‖‖‖‖‖	NWS, TLK	
AM	910	KALL	‖‖‖‖‖‖	NWS, TLK	
AM	1160	KSL	‖‖‖‖‖‖	NWS, TLK	
AM	1320	KFNZ	‖‖‖	SPRTS	

CASPER, WY

	FREQ	STATION	POWER	FORMAT	
	94.5	KMGW	‖‖‖‖‖‖	AC	
	95.5	KTRS	‖‖‖‖‖‖	TOP 40	
	106.9	KASS	‖‖‖‖‖‖	CLS RK	
AM	1230	KVOC	‖	OLDS	
	103.7	KQLT	‖‖‖‖‖‖	CTRY	
	104.7	KYOD	‖	CTRY	
	104.9	KZCY	‖	DVRS	
AM	1030	KTWO	‖‖‖‖‖‖	CTRY	
	90.3	KCSP	‖‖‖‖‖‖	RLG	
AM	830	KUYO	‖‖‖	RLG	

CHEYENNE, WY

	FREQ	STATION	POWER	FORMAT	
	106.3	KLEN	‖	AC	
	97.9	KIGN	‖‖‖‖‖‖	CLS RK	
	99.9	KRRR	‖	OLDS	
AM	1480	KRAE	‖	OLDS	
	100.7	KOLZ	‖‖‖	CTRY	

UT

88.1			
UT	SALT LAKE CITY	KWCR	105
88.3			
UT	SALT LAKE CITY	KCPW	105
88.5			
CO	DENVER	KGNU	102
88.7			
CO	COL. SPRINGS	KCME	102
89.1			
AZ	TUCSON	KUAZ	101
MT	MISSOULA	KUFM	104
UT	SALT LAKE CITY	KBYU	105
89.3			
CO	DENVER	KUVO	102
89.5			
AZ	PHOENIX	KBAQ	101
CO	GRAND JUNCT.	KPRN	102
CO	PUEBLO	KTSC	103
89.7			
CO	COL. SPRINGS	KEPC	102
89.9			
MT	GREAT FALLS	KGPR	103
MT	MISSOULA	KBGA	104
NM	ALBUQUERQUE	KUNM	104
90.1			
CO	DENVER	KCFR	102
UT	SALT LAKE CITY	KUER	105
90.3			
ID	BOISE	KBSU	103
WY	CASPER	KCSP	105
90.5			
AZ	TUCSON	KUAT	101
90.7			
NM	SANTA FE	KSFR	104
90.9			
UT	SALT LAKE CITY	KRCL	105
91.3			
AZ	TUCSON	KXCI	101
CO	GRAND JUNCT.	KMSA	102
91.5			
AZ	PHOENIX	KJZZ	101
CO	COL. SPRINGS	KRCC	102
ID	BOISE	KBSX	103
91.7			
MT	BILLINGS	KEMC	103
91.9			
CO	PUEBLO	KCFP	103
92.1			
AZ	TUCSON	KFMA	101
CO	DENVER	KJMN	102
92.3			
AZ	PHOENIX	KKFR	101
CO	GRAND JUNCT.	KJYE	102
ID	BOISE	KIZN	103
NM	ALBUQUERQUE	KRST	104
92.9			
AZ	TUCSON	KWFM	101
CO	COL. SPRINGS	KSPZ	102
MT	GREAT FALLS	KLFM	103
93.1			
CO	GRAND JUNCT.	KQIX	102
93.3			
AZ	PHOENIX	KDKB	101
MT	BILLINGS	KYYA	103
MT	MISSOULA	KGGL	104
NM	ALBUQUERQUE	KKOB	104
UT	SALT LAKE CITY	KUBL	105
93.7			
AZ	TUCSON	KRQQ	101
94.1			
ID	BOISE	KBXL	103
MT	BILLINGS	KRKX	103
NM	ALBUQUERQUE	KZRR	104
UT	SALT LAKE CITY	KODJ	105
94.3			
CO	COL. SPRINGS	KILO	102
94.5			
AZ	PHOENIX	KOOL	101
MT	GREAT FALLS	KMON	103
WY	CASPER	KMGW	105
94.7			
CO	DENVER	KRKS	102
94.9			
AZ	TUCSON	KMXZ	101
ID	BOISE	KFXJ	103
MT	MISSOULA	KYSS	104
UT	SALT LAKE CITY	KZHT	104
95.1			
CO	COL. SPRINGS	KRDO	102
95.5			
AZ	PHOENIX	KYOT	101
NM	SANTA FE	KMMG	104
WY	CASPER	KTRS	105
95.7			
CO	DENVER	KHIH	102
96.1			
AZ	TUCSON	KLPX	101
CO	COL. SPRINGS	KPRZ	102
96.3			
NM	ALBUQUERQUE	KHFM	104
UT	SALT LAKE CITY	KXRK	104
96.5			
CO	DENVER	KXPK	102
96.9			
AZ	PHOENIX	KMXP	101
CO	PUEBLO	KCCY	103
ID	BOISE	KKGL	103
97.1			
AZ	TUCSON	KGMS	101
MT	BILLINGS	KKBR	103
UT	SALT LAKE CITY	KISN	104
97.3			
CO	DENVER	KBCO	102
97.5			
AZ	TUCSON	KOAZ	101
97.7			
NM	ALBUQUERQUE	KLVO	104
97.9			
AZ	PHOENIX	KUPD	101
ID	BOISE	KQFC	103
UT	SALT LAKE CITY	KBZN	105
WY	CHEYENNE	KIGN	105
98.1			
CO	COL. SPRINGS	KKFM	102
98.3			
AZ	TUCSON	KOHT	101
98.5			
CO	DENVER	KYGO	102
MT	BILLINGS	KIDX	103
NM	SANTA FE	KABG	104
98.7			
AZ	PHOENIX	KKLT	101
UT	SALT LAKE CITY	KBEE	104
98.9			
MT	GREAT FALLS	KAAK	103
99.5			
AZ	TUCSON	KIIM	101
CO	DENVER	KKHK	102
NM	ALBUQUERQUE	KMGA	104
UT	SALT LAKE CITY	KURR	104
99.9			
AZ	PHOENIX	KESZ	101
CO	GRAND JUNCT.	KEKB	102
WY	CHEYENNE	KRRR	105
100.1			
MT	MISSOULA	KZOQ	103
100.3			
CO	DENVER	KIMN	102
NM	ALBUQUERQUE	KPEK	104
UT	SALT LAKE CITY	KSFI	104
100.7			
AZ	PHOENIX	KSLX	101
WY	CHEYENNE	KOLZ	105
101.1			
CO	DENVER	KOSI	102
NM	SANTA FE	KSFQ	104
UT	SALT LAKE CITY	KBER	104
101.3			
NM	ALBUQUERQUE	KEZF	104
101.5			
AZ	PHOENIX	KZON	101
101.7			
MT	BILLINGS	KRSQ	103
NM	ALBUQUERQUE	KZKL	104
101.89			
CO	COL. SPRINGS	KKCS	102
UT	SALT LAKE CITY	KKAT	105

WY	CHEYENNE	KMUS	105

102.5

AZ	PHOENIX	KNIX	101
MT	MISSOULA	KMSO	103
NM	ALBUQUERQUE	KIOT	104

102.7

CO	COL. SPRINGS	KBIQ	102
UT	SALT LAKE CITY	KQMB	104

102.9

MT	BILLINGS	KCTR	103

103.3

ID	BOISE	KARO	103
NM	ALBUQUERQUE	KTBL	104

103.5

AZ	PHOENIX	KWCY	101
CO	DENVER	KRFX	102
UT	SALT LAKE CITY	KRSP	104

103.7

MT	BILLINGS	KBBB	103
WY	CASPER	KQLT	105

104.1

AZ	TUCSON	KZPT	101

104.3

CO	DENVER	KCKK	102
CO	GRAND JUNCT.	KMXY	102
ID	BOISE	KLTB	103
UT	SALT LAKE CITY	KSOP	105

104.5

CO	PUEBLO	KYZX	103

104.7

AZ	PHOENIX	KZZP	101
NM	ALBUQUERQUE	KEXT	104
WY	CASPER	KYOD	105

104.9

WY	CHEYENNE	KZCY	105

105.1

CO	DENVER	KXKL	102
ID	BOISE	KJOT	103

105.3

AZ	PHOENIX	KMYL	101

105.5

CO	COL. SPRINGS	KSKX	102

105.7

UT	SALT LAKE CITY	KUMT	104

105.9

CO	DENVER	KALC	102
ID	BOISE	KCIX	103

106.1

MT	GREAT FALLS	KQDI	103

106.3

AZ	PHOENIX	KEDJ	101
CO	COL. SPRINGS	KKLI	102
NM	ALBUQUERQUE	KYLZ	104
WY	CHEYENNE	KLEN	105

106.5

UT	SALT LAKE CITY	KOSY	104

106.7

CO	DENVER	KBPI	102
NM	SANTA FE	KBOM	104

106.9

AZ	PHOENIX	KMJK	101
WY	CASPER	KASS	105

107.1

CO	PUEBLO	KNKN	103
ID	BOISE	KCID	103
NM	ALBUQUERQUE	KNKT	104

107.5

AZ	TUCSON	KHYT	101
CO	DENVER	KQKS	102
UT	SALT LAKE CITY	KENZ	104

107.9

AZ	PHOENIX	KMLE	101
CO	GRAND JUNCT.	KBKL	102
CO	PUEBLO	KDZA	103
ID	BOISE	KXLT	103
NM	ALBUQUERQUE	KTEG	104
UT	SALT LAKE CITY	KSNU	104

AM

550

AZ	PHOENIX	KOY	101

560

CO	DENVER	KLZ	102
MT	GREAT FALLS	KMON	103

570

UT	SALT LAKE CITY	KNRS	105

580

AZ	TUCSON	KSAZ	101
ID	BOISE	KFXD	103

590

CO	PUEBLO	KCSJ	103

610

NM	ALBUQUERQUE	KZSS	104

620

AZ	PHOENIX	KTAR	101

630

CO	DENVER	KHOW	102
ID	BOISE	KIDO	103
UT	SALT LAKE CITY	KTKK	105

650

WY	CHEYENNE	KGAB	105

670

ID	BOISE	KBOI	103

710

CO	DENVER	KNUS	102

730

ID	BOISE	KBSU	103
MT	BILLINGS	KURL	103

740

CO	COL. SPRINGS	KTWK	102

760

CO	DENVER	KTLK	102

770

NM	ALBUQUERQUE	KKOB	104

790

AZ	TUCSON	KNST	101
MT	BILLINGS	KGHL	103

830

AZ	TUCSON	KFLT	101
WY	CASPER	KUYO	105

850

CO	DENVER	KOA	102

860

AZ	PHOENIX	KMVP	101
NM	ALBUQUERQUE	KARS	104
UT	SALT LAKE CITY	KCNR	105

910

AZ	PHOENIX	KFYI	101
UT	SALT LAKE CITY	KALL	105

920

NM	ALBUQUERQUE	KHTL	104

930

MT	MISSOULA	KLCY	104

940

AZ	TUCSON	KCEE	101

950

CO	DENVER	KKFN	102

960

AZ	PHOENIX	KPXQ	101
UT	SALT LAKE CITY	KOVO	105

970

CO	PUEBLO	KFEL	103
MT	BILLINGS	KBUL	103

990

AZ	TUCSON	KTKT	101
CO	DENVER	KRKS	102

1000

NM	ALBUQUERQUE	KKIM	104

1030

AZ	TUCSON	KEVT	101
WY	CASPER	KTWO	105

1040

CO	COL. SPRINGS	KCBR	102

1050

NM	ALBUQUERQUE	KNML	104

1060

AZ	PHOENIX	KDUS	101
UT	SALT LAKE CITY	KKDS	105

1090

CO	DENVER	KMXA	102

1100

CO	GRAND JUNCT.	KNZZ	102

1140

ID	BOISE	KGEM	103

1150

CO	DENVER	KCUV	102
NM	ALBUQUERQUE	KDEF	104

1160

UT	SALT LAKE CITY	KSL	105

1190

AZ	PHOENIX	KMYL	101
NM	ALBUQUERQUE	KXKS	104

1210

AZ	TUCSON	KQTL	101

1230

AZ	PHOENIX	KISO	101
CO	GRAND JUNCT.	KEXO	102
CO	PUEBLO	KKPC	103
WY	CASPER	KVOC	105

1240

CO	COL. SPRINGS	KRDO	102
MT	BILLINGS	KMZK	103
NM	ALBUQUERQUE	KALY	104
WY	CHEYENNE	KFBC	105

1260

NM	SANTA FE	KVSF	104

1280

CO	DENVER	KRRF	102
UT	SALT LAKE CITY	KDYL	105

1290

AZ	TUCSON	KCUB	101
MT	MISSOULA	KGVO	104

1300

CO	COL. SPRINGS	KVOR	102

1310

AZ	PHOENIX	KXAM	101
MT	GREAT FALLS	KEIN	103
NM	ALBUQUERQUE	KIVA	104

1320

UT	SALT LAKE CITY	KFNZ	105

1340

CO	GRAND JUNCT.	KQIL	102
ID	BOISE	KTIK	103
MT	MISSOULA	KYLT	103

1350

CO	PUEBLO	KGHF	103
NM	ALBUQUERQUE	KABQ	104

1360

AZ	PHOENIX	KGME	101

1370

UT	SALT LAKE CITY	KSOP	105
WY	CHEYENNE	KJJL	105

1390

CO	DENVER	KJME	102

1400

AZ	PHOENIX	KSUN	101
AZ	TUCSON	KTUC	101
MT	GREAT FALLS	KXGF	103
NM	SANTA FE	KTRC	104
UT	SALT LAKE CITY	KSRR	105

1430

CO	DENVER	KEZW	102
UT	SALT LAKE CITY	KLO	105

1440

AZ	PHOENIX	KSLX	101

1450

AZ	TUCSON	KTZR	101
MT	GREAT FALLS	KQDI	103
MT	MISSOULA	KGRZ	104
NM	ALBUQUERQUE	KRZY	104

1460

CO	COL. SPRINGS	KKCS	102

1480

AZ	PHOENIX	KPHX	101
WY	CHEYENNE	KRAE	105

1490

AZ	TUCSON	KFFN	101
MT	BILLINGS	KBSR	103
NM	SANTA FE	KRSN	104

1510

AZ	PHOENIX	KFNN	101
CO	DENVER	KDKO	102

1550

AZ	TUCSON	KUAT	101

1580

AZ	PHOENIX	KCWW	101

1600

AZ	TUCSON	KXEW	101

E

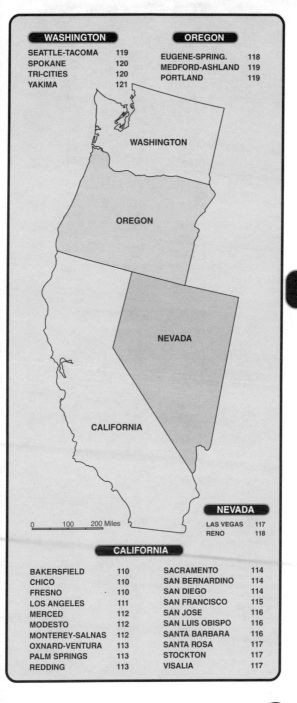

CLOSE-UP WEST COAST

F

	FREQ	STATION	POWER	FORMAT	
	95.3	KLLY	III	AC	
	99.3	KKBB	III	CLS RK	
	106.1	KRAB	III	ROCK	
	98.5	KSMJ	IIIII	OLDS	
	104.3	KCOO	III	OLDS	
	104.3	KISV	III	OLDS	
	104.5	KVLI	III	OLDS	
	105.3	KKDJ	IIIII	OLDS	
	102.5	KCNQ	III	CTRY	
	107.1	KCWR	I	CTRY	
	107.9	KUZZ	III	CTRY	
AM	550	KUZZ	IIIIIIII	CTRY	
	90.1	KTQX	I	SPAN	P
	92.1	KIWI	III	SPAN	
	97.7	KRME	III	SPAN	
	102.9	KSUV	III	SPAN	
	103.9	KMYX	III	SPAN	
AM	970	KAFY	III	SPAN	
AM	1010	KCHJ	IIIII	SPAN	
AM	1490	KWAC	I	SPAN	
	96.5	KKXX	IIIIIIII	JZZ	
	101.5	KGFM	III	EZ	
AM	1180	KERI	IIIIIIII	RLG	
AM	1230	KGEO	I	SPRTS	
AM	1410	KERN	I	NWS, TLK	
AM	1560	KNZR	IIIII	NWS, TLK	

			CHICO, CA		
	92.7	KLRS	III	HOT AC	
	95.1	KMXI	III	AC	
	101.5	KMJE	III	HOT AC	
	93.9	KFMF	III	ROCK	
	96.7	KZAP	III	CLS RK	
	103.5	KHSL	IIIIIIII	CTRY	
	97.7	KZCO	I	SPAN	
	90.1	KZFR	III	DVRS	P
	91.7	KCHO	III	CLS, NPR, PRI	P
	104.9	KYIX	I	RLG AC	
AM	1290	KPAY	IIIII	NWS, TLK	

			FRESNO, CA		
	101.1	KVSR	IIIIIIII	AC	
	102.7	KALZ	IIIII	HOT AC	
	94.3	KTAA	III	TOP 40	
	95.7	KJFX	III	CLS RK	
	97.9	KNAX	III	OLDS	
	93.7	KSKS	IIIIIIII	CTRY	
	91.5	KSJV	III	SPAN	P
	101.9	KOQO	IIIIIIII	SPAN	
	105.1	KLBN	IIIIIIII	SPAN	
	105.9	KRNC	III	SPAN	
AM	790	KOOR	IIIIIIII	SPAN	
AM	1430	KFIG	IIIII	SPAN	
AM	1550	KXEX	IIIII	SPAN	
AM	1600	KGST	IIIII	SPAN	
	96.7	KEZL	III	JZZ	
	99.3	KJWL	III	BB, NOS	
	88.1	KFCF	III	DVRS	P
	89.3	KVPR	III	CLS, NPR, PRI	P
	90.7	KFSR	I	JZZ	C
AM	580	KMJ	IIIIIIII	NWS, TLK	
AM	900	KBIF	III	BUS NWS	

CA

	FREQ	STATION	POWER	FORMAT	
AM	940	KFRE	IIIIIII	NWS, TLK	
AM	1300	KYNO	IIIII	SPRTS	
AM	1340	KCBL	I	SPRTS	

LOS ANGELES, CA

	FREQ	STATION	POWER	FORMAT	
	95.9	KEZY	III	AC	
	98.7	KYSR	IIIIIIII	HOT AC	
	103.5	KOST	IIIII	AC	
	104.3	KBIG	IIIIIIII	AC	
	106.3	KGMX	I	AC	
	102.7	KIIS	III	TOP 40	
	103.1	KACD	III	TOP 40	
	103.1	KBCD	III	TOP 40	
	105.9	KPWR	IIIIIIII	TOP 40	
	93.1	KCBS	IIIIIIII	CLS RK	
	95.5	KLOS	IIIIIIII	ROCK	
	106.7	KROQ	III	DVRS	
	107.1	KLYY	I	MOD RK	
	100.3	KCMG	III	OLDS	
	101.1	KRTH	IIIIIIII	OLDS	
AM	1110	KRLA	IIIIIIII	OLDS	
	93.9	KZLA	IIIII	CTRY	
	94.3	KIKF	I	CTRY	
	94.3	KBUA	I	SPAN	
	96.7	KWIZ	I	SPAN	
	97.9	KLAX	III	SPAN	
	101.9	KSCA	III	SPAN	
	105.5	KBUE	III	SPAN	
	107.5	KLVE	IIIIIIII	SPAN	
AM	900	KRRA	III	SPAN	
AM	930	KKHJ	IIIIIIII	SPAN	
AM	1020	KTNQ	IIIIIIII	SPAN	
AM	1330	KWKW	IIIII	SPAN	
	98.3	KRTO	III	BLK, R&B	
	103.9	KACE	III	BLK, R&B	
	94.7	KTWV	IIIIIIII	JZZ	
	105.1	KKGO	IIIII	CLS	
AM	570	KLAC	IIIIIIII	BB, NOS	
AM	1260	KGIL	IIIII	BB, NOS	
	88.1	KLON	III	JZZ, NPR	P
	88.5	KSBR	III	JZZ	C
	88.7	KSPC	I	AC	C
	88.9	KUCI	I	DVRS	C
	88.9	KXLU	I	DVRS	C
	89.3	KPCC	III	NWS, TLK, NPR, PRI	P
	89.9	KCRW	III	DVRS, NPR, PRI	P
	90.7	KPFK	IIIIIIII	DVRS	P
	91.5	KUSC	IIIII	CLS, NPR, PRI	P
	96.3	KFSG	IIIII	RLG AC	
AM	1230	KYPA	I	RLG AC	
	99.5	KKLA	IIIII	RLG	
	106.3	KALI	I	ETH	
	107.9	KWVE	IIIII	RLG	
AM	710	KDIS	IIIIIIII	KIDS	
AM	1300	KAZN	IIIII	ETH	
AM	1480	KVNR	IIIII	ETH	
	92.3	KKBT	IIIIIIII	URBAN	
	102.3	KJLH	III	URBAN	
	97.1	KLSX	IIIII	NWS, TLK	
AM	640	KFI	IIIIIIII	NWS, TLK	
AM	790	KABC	IIIIIIII	NWS, TLK	
AM	870	KIEV	IIIIIIII	NWS, TLK	

CA

CA

	FREQ	STATION	POWER	FORMAT	
AM	980	KFWB	IIIIIIII	NWS	
AM	1070	KNX	IIIIIIII	NWS	
AM	1150	KXTA	IIIIIIII	SPRTS	

MERCED, CA

	FREQ	STATION	POWER	FORMAT	
	104.7	KHTN	IIIII	AC	
	106.3	KIBG	III	AC	
	92.5	KVRQ	III	CLS RK	
	103.9	KDJK	III	CLS RK	
	97.5	KABX	IIIII	OLDS	
	96.3	KUBB	III	CTRY	
	106.9	KQLB	III	CTRY	
	98.7	KLOQ	III	SPAN	
	99.9	KCIV	III	RLG	
AM	1330	KLBS	I	ETH	
AM	1480	KYOS	IIIII	NWS, TLK	
AM	1580	KTFN	I	SPRTS	

MODESTO, CA

	FREQ	STATION	POWER	FORMAT	
	93.1	KOSO	IIIIIIII	HOT AC	
	102.3	KJSN	III	AC	
	98.3	KWNN	III	TOP 40	
	95.1	KHOP	IIIII	ROCK	
	104.1	KHKK	IIIII	CLS RK	
	103.3	KATM	IIIII	CTRY	
	88.7	KMPO	III	SPAN	P
	97.1	KZMS	I	SPAN	
AM	920	KLOC	III	SPAN	
AM	1390	KVIN	IIIII	BB, NOS	
	91.9	KCSS	I	DVRS	C
	89.5	KBES	I	ETH	P
AM	770	KCBC	IIIIIIII	RLG	
AM	860	KTRB	IIIIIIII	NWS, TLK	
AM	970	KANM	III	SPRTS	
AM	1360	KFIV	IIIII	NWS, TLK	

MONTEREY / SALINAS, CA

	FREQ	STATION	POWER	FORMAT	
	96.9	KWAV	IIIII	AC	
	102.1	KRKC	IIIIIIII	AC	
	106.3	KLUE	III	HOT AC	
	102.5	KDON	IIIII	TOP 40	
	92.7	KRQC	III	CLS RK	
	103.9	KISE	III	CLS RK	
	104.3	KMBY	III	ROCK	
	107.5	KPIG	III	ALT RK	
	105.1	KOCN	III	OLDS	
	100.7	KTOM	IIIIIIII	CTRY	
AM	1380	KTOM	IIIII	CTRY	
AM	1490	KRKC	I	CTRY	
	90.9	KHDC	I	SPAN	P
	93.9	KZSL	III	SPAN	
	97.9	KLXM	I	SPAN	
	99.1	KZOL	III	SPAN	
	99.5	KLOK	IIIII	SPAN	
	103.5	KRAY	I	SPAN	
	107.1	KVRG	III	SPAN	
AM	700	KVRG	IIIIIIII	SPAN	
AM	980	KCTY	III	SPAN	
AM	1570	KTGE	IIIII	SPAN	
	95.5	KBOQ	III	CLS	
	101.7	KXDC	III	JZZ	
AM	630	KIDD	IIIII	BB, NOS	

FREQ	STATION	POWER	FORMAT	
88.1	KZSC	III	DVRS	C
88.9	KUSP	III	CLS, NPR, PRI	P
90.3	KAZU	III	DVRS	P
AM 880	KKMC	IIIIIIII	RLG	
AM 540	KIEZ	IIIIIIII	NWS, TLK	
AM 1080	KSCO	IIIII	NWS, TLK	
AM 1240	KNRY	I	NWS, TLK	
AM 1340	KOMY	I	NWS, TLK	

OXNARD / VENTURA, CA

FREQ	STATION	POWER	FORMAT	
95.1	KBBY	III	AC	
105.5	KKBE	III	LGHT AC	
104.7	KCAQ	III	TOP 40	
95.9	KOCP	III	CLS RK	
107.1	KVYY	III	MOD RK	
100.7	KHAY	IIIIIIII	CTRY	
102.9	KXLM	I	SPAN	
103.7	KMLA	III	SPAN	
AM 910	KOXR	IIIIIIII	SPAN	
AM 1590	KXFS	IIIII	SPAN	
AM 1400	KKZZ	I	BB, NOS	
89.1	KCRU	I	ECLC, NPR, PRI	P
98.3	KDAR	III	RLG AC	
AM 1450	KVEN	I	NWS, TLK	
AM 1520	KTRO	IIIII	NWS, TLK	

PALM SPRINGS, CA

FREQ	STATION	POWER	FORMAT	
103.1	KEZN	III	AC	
92.7	KKUU	III	TOP 40	
100.5	KPSI	III	TOP 40	
93.7	KCLB	IIIII	MOD RK	
104.7	KDES	IIIII	OLDS	
106.1	KPLM	IIIII	CTRY	
94.7	KLOB	III	SPAN	
96.7	KUNA	III	SPAN	
AM 970	KCLB	IIIIIIII	SPAN	
98.5	KWXY	IIIII	BB, NOS	
102.3	KJJZ	I	JZZ	
AM 1140	KCMJ	IIIII	BB, NOS	
AM 1340	KWXY	I	BB, NOS	
88.9	KRTM	III	ALT RK	P
89.3	KCRY	III	DVRS, NPR, PRI	P
AM 920	KPSI	IIIIIIII	NWS, TLK	
AM 1010	KNWZ	III	NWS, TLK	
AM 1270	KXPS	IIIII	SPRTS	
AM 1400	KESQ	I	NWS, TLK	
AM 1450	KGAM	I	NWS, TLK	

REDDING, CA

FREQ	STATION	POWER	FORMAT	
104.3	KSHA	IIIIIIII	LGHT AC	
99.3	KNNN	III	TOP 40	
106.1	KRRX	IIIIIIII	MOD RK	
94.7	KEWB	IIIII	CTRY	
97.3	KNCQ	IIIIIIII	CTRY	
88.9	KFPR	IIIIIIII	CLS, NPR, PRI	P
89.7	KNCA	III	NWS, NPR, PRI	P
AM 600	KNRO	IIIII	SPRTS	
AM 1400	KQMS	I	NWS, TLK	

SACRAMENTO

FREQ	STATION	POWER	FORMAT	
92.5	KGBY	IIIII	AC	
96.1	KYMX	IIIII	AC	
102.5	KSFM	IIIII	TOP 40	
104.3	KQBR	III	TOP 40	

CA

FREQ	STATION	POWER	FORMAT	
107.9	KDND	IIIII	TOP 40	
93.9	KRLT	I	CLS RK	
96.9	KSEG	IIIII	CLS RK	
98.5	KRXQ	IIIII	ROCK	
100.5	KZZO	IIIII	DVRS	
106.5	KWOD	IIIII	DVRS	
101.1	KHYL	IIIII	OLDS	
AM 1470	KOME	IIIII	OLDS	
93.7	KRAK	IIIII	CTRY	
105.1	KNCI	IIIII	CTRY	
92.1	KZSA	III	SPAN	
97.9	KTTA	III	SPAN	
101.9	KRRE	IIIIIIII	SPAN	
AM 1240	KSQR	I	SPAN	
AM 1320	KCTC	IIIII	BB, NOS	
88.9	KXJZ	IIIII	JZZ, NPR, PRI	P
90.3	KDVS	I	DVRS	C
90.9	KXPR	IIIII	CLS, NPR, PRI	P
AM 710	KFIA	IIIIIIII	RLG AC	
AM 1380	KTKZ	IIIII	RLG AC	
103.5	KBMB	I	URBAN	
AM 650	KSTE	IIIIIIII	NWS, TLK	
AM 1140	KHTK	IIIIIIII	NWS, TLK	
AM 1530	KFBK	IIIIIIII	NWS, TLK	

SAN BERNARDINO, CA

FREQ	STATION	POWER	FORMAT	
92.7	KELT	III	LGHT AC	
101.3	KATY	III	AC	
AM 1290	KKDD	IIIII	AC	
99.1	KGGI	III	TOP 40	
96.7	KCAL	III	CLS RK	
103.9	KCXX	III	DVRS	
99.9	KOLA	III	OLDS	
95.1	KFRG	IIIII	CTRY	
AM 1350	KCKC	IIIII	CTRY	
97.5	KSSE	IIIIIIII	SPAN	
101.7	KXSB	III	SPAN	
105.7	KXRS	III	SPAN	
AM 590	KSZZ	IIIII	SPAN	
AM 1410	KCAL	IIIII	SPAN	
AM 1440	KDIF	I	SPAN	
96.1	KWRP	III	BB, NOS	
88.3	KUCR	III	DVRS	C
91.9	KVCR	III	CLS, NPR, PRI	P
89.7	KSGN	III	RLG	P
AM 1240	KKLA	I	RLG	
AM 1370	KWRM	IIIII	ETH	

SAN DIEGO, CA

FREQ	STATION	POWER	FORMAT	
94.1	KCDE	III	LGHT AC	
94.1	KJQY	IIIIIIII	LGHT AC	
96.5	KYXY	IIIII	AC	
100.7	KFMB	IIIII	AC	
90.3	XHTZ	IIIII	TOP 40	
93.3	KHTS	III	TOP 40	
101.5	KGB	IIIIIIII	CLS RK	
102.1	KXST	IIIII	ALT RK	
103.7	KPLN	IIIII	CLS RK	
105.3	KIOZ	IIIII	ROCK	
107.1	KSYY	III	MOD RK	
AM 1320	KKSM	I	DVRS	
94.9	KBZT	III	OLDS	

CA

	FREQ	STATION	POWER	FORMAT	
	97.3	KSON	IIIII	CTRY	
	99.3	XHKY	III	SPAN	
	104.5	XLTN	IIIII	SPAN	
	106.5	KKLQ	IIIII	SPAN	
AM	1040	KURS	IIIII	SPAN	
	95.7	KMCG	III	BLK, R&B	
	92.1	KFSD	III	CLS	
	98.1	KIFM	IIIII	JZZ	
AM	1360	KPOP	IIIII	BB, NOS	
AM	1450	KSPA	I	BB, NOS	
	88.3	KSDS	I	JZZ	C
	89.5	KPBS	III	NWS, TLK, NPR, PRI	P
AM	1210	KPRZ	IIIII	RLG	
AM	1240	KSON	I	KIDS	
AM	600	KOGO	IIIIIIII	NWS, TLK	
AM	760	KFMB	IIIIIIII	NWS, TLK	
AM	1000	KCEO	IIIII	BUS NWS	
AM	1130	KSDO	IIIII	NWS, TLK	
AM	1170	KCBQ	IIIIIIII	NWS, TLK	
	SAN FRANCISCO, CA				
	95.3	KUIC	III	AC	
	96.5	KOIT	IIIIIIII	AC	
	97.3	KLLC	IIIIIIII	AC	
	98.1	KISQ	IIIIIIII	AC	
	101.3	KIOI	IIIIIIII	HOT AC	
	102.9	KBLX	III	AC	
AM	910	KNEW	IIIIIIII	HOT AC	
AM	1260	KOIT	IIIII	AC	
	94.9	KYLD	III	TOP 40	
	95.7	KZQZ	III	TOP 40	
	106.1	KMEL	IIIIIIII	TOP 40	
	92.1	KFJO	I	ROCK	
	104.5	KFOG	III	ALT RK	
	105.3	KITS	IIIII	MOD RK	
	107.7	KSAN	IIIIIIII	MOD RK	
	99.7	KFRC	IIIIIIII	OLDS	
AM	610	KFRC	IIIIIIII	OLDS	
	93.3	KYCY	IIIII	CTRY	
AM	1550	KYCY	IIIII	CTRY	
	98.9	KSOL	III	SPAN	
AM	1010	KIQI	IIIII	SPAN	
	100.7	KKHI	III	CLS	
	102.1	KDFC	IIIIIIII	CLS	
	103.7	KKSF	III	JZZ	
AM	960	KABL	IIIIIIII	BB, NOS	
	88.5	KQED	IIIIIIII	NWS, TLK, NPR, PRI	P
	89.3	KPFB	I	DVRS	P
	89.5	KSMC	I	DVRS	C
	89.5	KPOO	III	BLK, R&B	P
	90.3	KUSF	III	DVRS	C
	90.7	KALX	III	DVRS	C
	91.1	KCSM	III	JZZ, NPR, PRI	P
	94.1	KPFA	IIIIIIII	DVRS	P
AM	1100	KFAX	IIIIIIII	RLG AC	
	89.9	KNDL	III	RLG	C
AM	1310	KMKY	IIIII	KIDS	
AM	1400	KVTO	I	ETH	
AM	1450	KEST	I	ETH	
AM	560	KSFO	IIIIIIII	NWS, TLK	
AM	680	KNBR	IIIIIIII	SPRTS	
AM	740	KCBS	IIIIIIII	NWS	
AM	810	KGO	IIIIIIII	NWS, TLK	
AM	1050	KTCT	IIIIIIII	SPRTS	

CA

FREQ	STATION	POWER	FORMAT	
94.5	KBAY	IIIIIIII	LGHT AC	
101.7	KKIQ	III	AC	
105.7	KARA	IIIII	AC	
106.5	KEZR	IIIII	AC	
92.3	KSJO	IIIII	ROCK	
97.7	KFFG	III	DVRS	
98.5	KUFX	III	CLS RK	
104.9	KLDZ	IIIII	CLS RK	
95.3	KRTY	III	CTRY	
100.3	KBRG	IIIIIIII	SPAN	
AM 1170	KLOK	IIIIIIII	SPAN	
AM 1370	KZSF	IIIII	SPAN	
89.3	KOHL	III	TOP 40	C
89.7	KFJC	III	AC	C
90.1	KZSU	I	DVRS	C
90.5	KSJS	III	JZZ	C
91.5	KKUP	III	DVRS	P
AM 1430	KVVN	I	ETH	
AM 1500	KSJX	IIIII	ETH	
AM 1220	KBPA	IIIII	NWS, TLK	
AM 1590	KLIV	IIIII	NWS	

SAN LUIS OBISPO, CA

FREQ	STATION	POWER	FORMAT	
101.3	KSTT	IIIIIIII	AC	
103.1	KBZX	III	AC	
96.1	KSLY	IIIIIIII	TOP 40	
99.7	KWWV	III	TOP 40	
93.3	KZOZ	IIIII	CLS RK	
94.9	KOTR	III	ALT RK	
95.3	KXTZ	III	CLS RK	
104.5	KIQO	III	OLDS	
92.5	KDDB	III	CTRY	
98.1	KKJG	IIIIIIII	CTRY	
106.1	KWEZ	III	CTRY	
AM 1030	KJDJ	III	SPAN	
107.3	KQJZ	III	JZZ	
90.1	KCBX	III	CLS, NPR, PRI	P
91.3	KCPR	I	DVRS	C
97.1	KWQH	III	RLG AC	
AM 920	KVEC	III	NWS, TLK	
AM 1230	KPRL	I	NWS, TLK	
AM 1280	KKAL	IIIII	SPRTS	
AM 1340	KGLW	I	NWS, TLK	

SANTA BARBARA, CA

FREQ	STATION	POWER	FORMAT	
101.7	KSBL	III	AC	
103.3	KRUZ	IIIIIIII	AC	
92.9	KJEE	III	MOD RK	
99.9	KTYD	IIIIIIII	CLS RK	
106.3	KHTY	IIIII	CLS RK	
94.5	KSPE	III	SPAN	
AM 1490	KBKO	I	SPAN	
93.7	KDB	III	CLS	
106.3	KMGQ	III	JZZ	
AM 1290	KZBN	I	BB, NOS	
91.9	KCSB	III	DVRS	C
AM 990	KTMS	IIIIIIII	NWS, TLK	
AM 1250	KEYT	III	NWS, TLK	
AM 1340	KXXT	I	SPRTS	

CA

	FREQ	STATION	POWER	FORMAT	
	95.9	KHBG	I	AC	
	100.1	KZST	I	AC	
	104.1	KMHX	III	AC	
	101.7	KXFX	III	ROCK	
	97.7	KMGG	III	OLDS	
	92.9	KFGY	IIIIIIII	CTRY	
	104.9	KRPQ	III	CTRY	
	89.1	KBBF	III	SPAN	P
AM	1460	KRRS	I	SPAN	
	93.7	KJZY	III	JZZ	
	91.1	KRCB	I	CLS, NPR, PRI	P
AM	1350	KSRO	IIIII	NWS, TLK	

STOCKTON, CA

	FREQ	STATION	POWER	FORMAT	
	99.3	KJOY	III	AC	
	97.7	KWIN	III	TOP 40	
AM	1420	KSTN	IIIII	TOP 40	
	100.1	KQOD	III	OLDS	
	100.9	KMIX	III	SPAN	
	107.3	KSTN	III	SPAN	
	89.5	KJQN	III	DVRS	C
	91.3	KUOP	III	JZZ, CLS, NPR, PRI	P
AM	1280	KJAX	I	NWS, TLK	

VISALIA, CA

	FREQ	STATION	POWER	FORMAT	
	98.9	KSOF	III	LGHT AC	
	94.9	KBOS	IIIIIIII	TOP 40	
	97.1	KSEQ	III	TOP 40	
	99.7	KIOO	III	CLS RK	
	103.7	KRZR	IIIII	MOD RK	
	104.1	KFRR	III	DVRS	
	92.9	KFSO	IIIIIIII	OLDS	
	104.9	KCRZ	III	OLDS	
AM	1400	KFSO	I	OLDS	
	106.7	KJUG	III	CTRY	
AM	1270	KJUG	IIIII	CTRY	
	90.5	KUFW	III	SPAN	P
	94.5	KGEN	I	SPAN	
	100.5	KOJJ	III	SPAN	
AM	1240	KJOP	I	SPAN	
AM	1370	KGEN	I	SPAN	
	96.1	KSLK	III	JZZ	
	107.5	KMPH	IIIIIIII	BUS NWS	
AM	1450	KTIP	I	NWS, TLK	

LAS VEGAS, NV

	FREQ	STATION	POWER	FORMAT	
	94.1	KMXB	IIIIIIII	HOT AC	
	100.5	KMZQ	IIIIIIII	AC	
	105.5	KSTJ	IIIIIIII	HOT AC	
	106.5	KSNE	IIIIIIII	AC	
	98.5	KLUC	IIIIIIII	TOP 40	
	92.3	KOMP	IIIIIIII	ROCK	
	96.3	KKLZ	IIIIIIII	CLS RK	
	97.1	KXPT	IIIIIIII	ALT RK	
	93.1	KQOL	IIIIIIII	OLDS	
	95.5	KWNR	IIIIIIII	CTRY	
	101.9	KFMS	IIIIIIII	CTRY	
AM	870	KLSQ	IIIIIIII	SPAN	
	103.5	KISF	IIIIIIII	BLK, R&B	
	104.3	KJUL	IIIIIIII	BB, NOS	
	89.5	KNPR	IIIIIIII	CLS, NPR, PRI	P

CA
NV

	FREQ	STATION	POWER	FORMAT	
	91.5	KUNV	IIIII	JZZ, NPR	C
	88.1	KCEP	III	URBAN	P
	105.1	KVBC	IIIIIIIII	NWS, TLK	
AM	720	KDWN	IIIIIIIII	NWS, TLK	
AM	840	KXNF	IIIIIIIII	NWS, TLK	
AM	920	KBAD	IIIIIIIII	SPRTS	
AM	970	KNUU	IIIIIIIII	NWS, TLK	
AM	1230	KLAV	I	NWS, TLK	
AM	1340	KRLV	I	NWS, TLK	
AM	1460	KENO	IIIII	SPRTS	

RENO, NV

	FREQ	STATION	POWER	FORMAT	
	95.5	KNEV	IIIIIIIII	AC	
	106.9	KRNO	IIIIIIIII	AC	
	97.3	KWNZ	IIIIIIIII	TOP 40	
	92.9	KNHK	IIIIIIIII	CLS RK	
	100.1	KTHX	III	ALT RK	
	100.9	KRZQ	III	DVRS	
	104.5	KDOT	IIIIIIIII	MOD RK	
	105.7	KOZZ	IIIIIIIII	CLS RK	
AM	1300	KPTL	IIIII	OLDS	
	98.1	KBUL	IIIIIIIII	CTRY	
AM	1450	KHIT	IIIII	CTRY	
AM	920	KQLO	IIIIIIIII	SPAN	
AM	1340	KXEQ	I	SPAN	
	92.1	KSRN	III	BB, NOS	
	94.7	KSSJ	IIIIIIIII	JZZ	
AM	1230	KCBN	I	BB, NOS	
	88.7	KUNR	IIIII	CLS, NPR, PRI	P
	101.7	KRNV	IIIII	NWS, TLK	
AM	630	KPTT	IIIIIIIII	NWS, TLK	
AM	780	KKOH	IIIIIIIII	NWS, TLK	
AM	1270	KPLY	IIIII	SPRTS	

EUGENE / SPRING., OR

	FREQ	STATION	POWER	FORMAT	
	94.5	KMGE	IIIIIIIII	AC	
	106.9	KCST	III	LGHT AC	
	104.7	KDUK	IIIIIIIII	TOP 40	
	95.3	KNRQ	III	DVRS	
	96.1	KZEL	IIIIIIIII	CLS RK	
	99.1	KODZ	IIIIIIIII	OLDS	
AM	840	KKNX	III	OLDS	
	93.1	KKNU	IIIIIIIII	CTRY	
	97.9	KKTT	IIIIIIIII	CTRY	
AM	1400	KNND	I	CTRY	
AM	1450	KKXO	I	BB, NOS	
	88.1	KWVA	I	DVRS	C
	89.7	KLCC	IIIIIIIII	JZZ, NPR	P
AM	1050	KORE	IIIII	RLG	
AM	590	KUGN	IIIIIIIII	NWS, TLK	
AM	1120	KPNW	IIIIIIIII	NWS, TLK	

MEDFORD/ASHLAND, OR

	FREQ	STATION	POWER	FORMAT	
	101.9	KCMX	IIIIIIIII	AC	
	107.5	KKJJ	III	AC	
	93.7	KTMT	IIIIIIIII	TOP 40	
	95.7	KBOY	IIIIIIIII	CLS RK	
	106.3	KZZE	III	MOD RK	
	94.7	KRRM	III	CTRY	
	100.3	KRWQ	IIIII	CTRY	
	105.1	KAKT	IIIIIIIII	CTRY	
AM	610	KRTA	IIIIIIIII	SPAN	

NV OR

	FREQ	STATION	POWER	FORMAT	
AM	1440	KMED	IIIII	BB, NOS	
	88.3	KSRG	I	NWS, NPR, PRI	P
	89.1	KSMF	III	NWS, NPR, PRI	P
	90.1	KSOR	IIIIIIII	NWS, NPR	P
AM	1230	KSJK	I	NWS, TLK, PRI	P
	103.5	KOPE	IIIIIIII	NWS, TLK	
AM	580	KCMX	IIIII	NWS, TLK	
AM	880	KTMT	III	SPRTS	

PORTLAND, OR

	FREQ	STATION	POWER	FORMAT	
	95.5	KXL	IIIIIIII	AC	
	103.3	KKCW	IIIIIIII	AC	
	105.1	KRSK	IIIIIIII	HOT AC	
	107.5	KBBT	III	HOT AC	
	100.3	KKRZ	IIIIIIII	TOP 40	
	92.3	KGON	IIIIIIII	CLS RK	
	94.7	KNRK	III	MOD RK	
	101.1	KUFO	IIIIIIII	ROCK	
	101.9	KINK	IIIIIIII	ALT RK	
	97.1	KKSN	IIIIIIII	OLDS	
	98.7	KUPL	IIIIIIII	CTRY	
	99.5	KWJJ	IIIIIIII	CTRY	
AM	970	KUPL	IIIIIIII	CTRY	
AM	940	KWBY	III	SPAN	
AM	1230	KMUZ	I	SPAN	
	106.7	KKJZ	IIIIIIII	JZZ	
	89.1	KMHD	III	JZZ	C
	89.9	KBPS	III	CLS, NPR, PRI	P
	90.3	KSLC	I	AC	C
	90.7	KBOO	IIII	DVRS	P
	91.5	KOPB	IIIIIIII	NWS, NPR, PRI	P
	93.7	KPDQ	IIIIIIII	RLG AC	
AM	800	KPDQ	III	RLG AC	
AM	1330	KKPZ	IIII	RLG	
AM	1480	KBMS	I	URBAN	
AM	620	KEWS	IIIIIIII	NWS, TLK	
AM	750	KXL	IIIIIIII	NWS, TLK	
AM	910	KFXX	IIIIIIII	SPRTS	
AM	910	KKSN	IIIIIIII	SPRTS	
AM	1080	KOTK	IIIIIIII	NWS, TLK	
AM	1190	KEX	IIIIIIII	NWS, TLK	
AM	1360	KUIK	IIII	NWS, TLK	
AM	1430	KYKN	IIII	NWS, TLK	
AM	1550	KVAN	IIIII	NWS, TLK	

SEATTLE / TACOMA, WA

	FREQ	STATION	POWER	FORMAT	
	92.5	KLSY	IIIIIIII	AC	
	96.1	KXXO	IIIIIIII	AC	
	101.5	KPLZ	IIIIIIII	AC	
	106.1	KBKS	IIIIIIII	HOT AC	
	106.9	KRWM	IIIIIIII	LGHT AC	
AM	1090	KRPM	IIIIIIII	HOT AC	
AM	1240	KGY	I	AC	
	93.3	KUBE	IIIIIIII	TOP 40	
	95.7	KJR	IIIIIIII	CLS RK	
	99.9	KISW	IIIIIIII	ROCK	
	102.5	KZOK	IIIIIIII	CLS RK	
	103.7	KMTT	IIIIIIII	ALT RK	
	107.7	KNDD	IIIIIIII	MOD RK	
	97.3	KBSG	IIIIIIII	OLDS	
AM	1210	KBSG	IIIII	OLDS	
	94.1	KMPS	IIIIIIII	CTRY	

OR
WA

119

FREQ	STATION	POWER	FORMAT	
96.5	KYCW	IIIIIIII	CTRY	
98.1	KING	IIIIIIII	CLS	
98.9	KWJZ	IIIIIIII	JZZ	
AM 880	KIXI	IIIIIIII	BB, NOS	
88.5	KPLU	IIIIIIII	JZZ, NPR, PRI	P
89.3	KAOS	I	DVRS, PRI	P
89.9	KGRG	I	MOD RK	C
90.1	KUPS	I	AC	C
90.3	KCMU	III	ALT, PRI	C
90.7	KSER	III	DVRS, PRI	P
90.9	KVTI	IIIII	TOP 40	C
91.3	KBCS	I	DVRS, PRI	C
94.9	KUOW	IIIIIIII	NWS, TLK, NPR, PRI	P
105.3	KCMS	IIIIIIII	RLG AC	
AM 630	KCIS	IIIIIIII	RLG AC	
AM 820	KGNW	IIIIIIII	RLG	
AM 1300	KKOL	IIIII	RLG	
104.9	KKBY	III	URBAN	
AM 1420	KRIZ	I	URBAN	
100.7	KIRO	IIIIIIII	NWS, TLK	
AM 570	KVI	IIIIIIII	NWS, TLK	
AM 710	KIRO	IIIIIIII	NWS, TLK	
AM 770	KNWX	IIIIIIII	NWS	
AM 850	KHHO	IIIIIIII	NWS, TLK	
AM 950	KJR	IIIIIIII	SPRTS	
AM 1000	KOMO	IIIIIIII	NWS, TLK	

SPOKANE, WA

FREQ	STATION	POWER	FORMAT	
98.1	KISC	IIIIIIII	AC	
99.9	KXLY	IIIIIIII	LGHT AC	
AM 1080	KVNI	IIIII	AC	
92.9	KZZU	IIIIIIII	TOP 40	
94.5	KHTQ	IIIIIIII	TOP 40	
98.9	KKZX	IIIIIIII	CLS RK	
103.9	KNJY	III	DVRS	
105.7	KAEP	IIIIIIII	DVRS	
101.1	KEYF	IIIIIIII	OLDS	
AM 1050	KEYF	IIIII	OLDS	
93.7	KDRK	IIIIIIII	CTRY	
96.1	KNFR	IIIIIIII	CTRY	
103.1	KCDA	IIIIIIII	CTRY	
AM 590	KAQQ	IIIIIIII	BB, NOS	
89.5	KEWU	IIIII	JZZ	C
91.1	KPBX	IIIIIIII	CLS, NPR, PRI	P
91.9	KSFC	I	CLS, NPR, PRI	C
101.9	KTSL	IIIII	RLG AC	
AM 790	KJRB	IIIIIIII	SPRTS	
AM 920	KXLY	IIIIIIII	NWS, TLK	
AM 970	KTRW	IIIIIIII	SPRTS	
AM 1510	KGA	IIIIIIII	NWS, TLK	

TRI-CITIES, WA

FREQ	STATION	POWER	FORMAT	
98.3	KEYW	I	HOT AC	
101.7	KZXR	IIIII	HOT AC	
105.3	KONA	IIIIIIII	LGHT AC	
AM 610	KONA	IIIIIIII	AC	
106.5	KEGX	IIIIIIII	CLS RK	
102.7	KORD	IIIIIIII	CTRY	
AM 960	KALE	IIIIIIII	BB, NOS	
89.1	KFAE	IIIIIIII	CLS, NPR, PRI	P
101.3	KGDN	III	RLG	
AM 870	KFLD	IIIIIIII	SPRTS	

	FREQ	STATION	POWER	FORMAT	
AM	1340	KTCR	I	NWS, TLK	

YAKIMA, WA

	FREQ	STATION	POWER	FORMAT	
	92.9	KQSN	III	AC	
	105.7	KRSE	IIIIIIIII	AC	
	107.3	KFFM	IIIIIIIII	TOP 40	
	94.5	KATS	IIIIIIIII	CLS RK	
	99.7	KHHK	IIIIIIIII	CLS RK	
AM	980	KJOX	IIIIIIIII	CLS RK	
AM	1460	KMWX	IIIII	OLDS	
	100.9	KARY	I	CTRY	
	104.1	KXDD	IIIIIIIII	CTRY	
AM	1490	KENE	I	CTRY	
	96.7	KZTB	I	SPAN	
	96.9	KZTA	I	SPAN	
AM	1020	KYXE	IIIII	SPAN	
AM	1210	KREW	IIIII	BB, NOS	
	90.3	KNWY	III	NWS, NPR, PRI	P
AM	1390	KBBO	IIIII	RLG	
AM	1280	KIT	IIIII	NWS, TLK	

WA

		88.1	
CA	FRESNO	KFCF	110
CA	LOS ANGELES	KLON	111
CA	MONTEREY-SALIN.	KZSC	113
NV	LAS VEGAS	KCEP	118
OR	EUGENE	KWVA	118

		88.3	
CA	SAN BERNARD.	KUCR	114
CA	SAN DIEGO	KSDS	115
OR	MEDFORD	KSRG	119
CA	LOS ANGELES	KSBR	111
CA	SAN FRANCISCO	KQED	116
WA	SEATTLE-TACOM	KPLU	120

		88.7	
CA	LOS ANGELES	KSPC	111
CA	MODESTO	KMPO	112
NV	RENO	KUNR	118

		88.9	
CA	LOS ANGELES	KUCI	111
CA	LOS ANGELES	KXLU	111
CA	MONTEREY-SALIN.	KUSP	113
CA	PALM SPRINGS	KRTM	113
CA	REDDING	KFPR	113
CA	SACRAMENTO	KXJZ	114

		89.1	
CA	OXNARD-VENTUR.	KCRU	113
CA	SANTA ROSA	KBBF	117
OR	MEDFORD	KSMF	119
OR	PORTLAND	KMHD	119
WA	RICH-PASCO-KEN	KFAE	121

		89.3	
CA	FRESNO	KVPR	110
CA	LOS ANGELES	KPCC	111
CA	PALM SPRINGS	KCRY	113
CA	SAN FRANCISCO	KOHL	116
CA	SAN FRANCISCO	KPFB	116
WA	SEATTLE-TACOM	KAOS	120

		89.5	
CA	MODESTO	KBES	112
CA	SAN DIEGO	KPBS	115
CA	SAN FRANCISCO	KSMC	116
CA	SAN FRANCISCO	KPOO	116
CA	STOCKTON	KJQN	117
NV	LAS VEGAS	KNPR	118
WA	SPOKANE	KEWU	120

		89.7	
CA	REDDING	KNCA	113
CA	SAN BERNARD.	KSGN	114
CA	SAN FRANCISCO	KFJC	116
OR	EUGENE	KLCC	118

		89.9	
CA	LOS ANGELES	KCRW	111
CA	SAN FRANCISCO	KNDL	116
OR	PORTLAND	KBPS	119
WA	SEATTLE-TACOM	KGRG	120

		90.1	
CA	BAKERSFIELD	KTQX	110
CA	CHICO	KZFR	110
CA	SAN FRANCISCO	KZSU	116
CA	SAN LUIS OBIS.	KCBX	116
OR	MEDFORD	KSOR	119
WA	SEATTLE-TACOM	KUPS	120

		90.3	
CA	MONTEREY-SALIN.	KAZU	113
CA	SACRAMENTO	KDVS	114
CA	SAN DIEGO	XHTZ	114
CA	SAN FRANCISCO	KUSF	116
OR	PORTLAND	KSLC	119
WA	SEATTLE-TACOM	KCMU	120
WA	YAKIMA	KNWY	121

		90.5	
CA	SAN FRANCISCO	KSJS	116
CA	VISALIA	KUFW	117

		90.7	
CA	FRESNO	KFSR	110
CA	LOS ANGELES	KPFK	111
CA	SAN FRANCISCO	KALX	116
OR	PORTLAND	KBOO	119
WA	SEATTLE-TACOM	KSER	120

		90.9	
CA	MONTEREY-SALIN.	KHDC	112
CA	SACRAMENTO	KXPR	114
WA	SEATTLE-TACOM	KVTI	120

		91.1	
CA	SAN FRANCISCO	KCSM	116
CA	SANTA ROSA	KRCB	117
WA	SPOKANE	KPBX	120

		91.3	
CA	SAN LUIS OBIS.	KCPR	116
CA	STOCKTON	KUOP	117
WA	SEATTLE-TACOM	KBCS	120

		91.5	
CA	FRESNO	KSJV	110
CA	LOS ANGELES	KUSC	111
CA	SAN FRANCISCO	KKUP	116
NV	LAS VEGAS	KUNV	118
OR	PORTLAND	KOPB	119

		91.7	
CA	CHICO	KCHO	110

		91.9	
CA	MODESTO	KCSS	112
CA	SAN BERNARD.	KVCR	114
CA	SANTA BARBARA	KCSB	116
WA	SPOKANE	KSFC	120

		92.1	
CA	BAKERSFIELD	KIWI	110
CA	SACRAMENTO	KZSA	114
CA	SAN DIEGO	KFSD	115
CA	SAN FRANCISCO	KFJO	115
NV	RENO	KSRN	118

		92.3	
CA	LOS ANGELES	KKBT	111
CA	SAN FRANCISCO	KSJO	115
NV	LAS VEGAS	KOMP	117
OR	PORTLAND	KGON	119

		92.5	
CA	MERCED	KVRQ	112
CA	SACRAMENTO	KGBY	114
CA	SAN LUIS OBIS.	KDDB	116
WA	SEATTLE-TACOM	KLSY	119

		92.7	
CA	CHICO	KLRS	110
CA	MONTEREY-SALIN.	KRQC	112
CA	PALM SPRINGS	KKUU	113
CA	SAN BERNARD.	KELT	114

		92.9	
CA	SANTA BARBARA	KJEE	116
CA	SANTA ROSA	KFGY	117
CA	VISALIA	KFSO	117
NV	RENO	KNHK	118
WA	SPOKANE	KZZU	120
WA	YAKIMA	KQSN	121

		93.1	
CA	LOS ANGELES	KCBS	111
CA	MODESTO	KOSO	112
NV	LAS VEGAS	KQOL	117
OR	EUGENE	KKNU	118

		93.3	
CA	SAN DIEGO	KHTS	114
CA	SAN FRANCISCO	KYCY	115
CA	SAN LUIS OBIS.	KZOZ	116
WA	SEATTLE-TACOM	KUBE	119

		93.7	
CA	FRESNO	KSKS	110
CA	PALM SPRINGS	KCLB	113
CA	SACRAMENTO	KRAK	114
CA	SANTA BARBARA	KDB	116
CA	SANTA ROSA	KJZY	117
OR	MEDFORD	KTMT	118
OR	PORTLAND	KPDQ	119
WA	SPOKANE	KDRK	120

		93.9	
CA	CHICO	KFMF	110
CA	LOS ANGELES	KZLA	111
CA	MONTEREY-SALIN.	KZSL	112
CA	SACRAMENTO	KRLT	114

		94.1	
CA	SAN DIEGO	KCDE	114
CA	SAN DIEGO	KJQY	114
CA	SAN FRANCISCO	KPFA	116
NV	LAS VEGAS	KMXB	117
WA	SEATTLE-TACOM	KMPS	120

		94.3	
CA	FRESNO	KTAA	110
CA	LOS ANGELES	KIKF	111
CA	LOS ANGELES	KBUA	111

94.5

CA	SAN FRANCISCO	KBAY	115
CA	SANTA BARBARA	KSPE	116
CA	VISALIA	KGEN	117
OR	EUGENE	KMGE	118
WA	SPOKANE	KHTQ	120
WA	YAKIMA	KATS	121

94.7

CA	LOS ANGELES	KTWV	111
CA	PALM SPRINGS	KLOB	113
CA	REDDING	KEWB	113
NV	RENO	KSSJ	118
OR	MEDFORD	KRRM	119
OR	PORTLAND	KNRK	119

94.9

CA	SAN DIEGO	KBZT	115
CA	SAN FRANCISCO	KYLD	115
CA	SAN LUIS OBIS.	KOTR	116
CA	VISALIA	KBOS	117
WA	SEATTLE-TACOM	KUOW	120

95.1

CA	CHICO	KMXI	110
CA	MODESTO	KHOP	112
CA	OXNARD-VENTUR.	KBBY	113
CA	SAN BERNARD.	KFRG	114

95.3

CA	BAKERSFIELD	KLLY	110
CA	SAN FRANCISCO	KUIC	115
CA	SAN FRANCISCO	KRTY	115
CA	SAN LUIS OBIS.	KXTZ	116
OR	EUGENE	KNRQ	118

95.5

CA	LOS ANGELES	KLOS	111
CA	MONTEREY-SALIN.	KBOQ	112
NV	LAS VEGAS	KWNR	117
NV	RENO	KNEV	118
OR	PORTLAND	KXL	119

95.7

CA	FRESNO	KJFX	110
CA	SAN DIEGO	KMCG	115
CA	SAN FRANCISCO	KZQZ	115
OR	MEDFORD	KBOY	118
WA	SEATTLE-TACOM	KJR	119

95.9

CA	LOS ANGELES	KEZY	111
CA	OXNARD-VENTUR.	KOCP	113
CA	SANTA ROSA	KHBG	117

96.1

CA	SACRAMENTO	KYMX	114
CA	SAN BERNARD.	KWRP	114
CA	SAN LUIS OBIS.	KSLY	116
CA	VISALIA	KSLK	117
OR	EUGENE	KZEL	118
WA	SEATTLE-TACOM	KXXO	119
WA	SPOKANE	KNFR	120

96.3

CA	LOS ANGELES	KFSG	111
CA	MERCED	KUBB	112
NV	LAS VEGAS	KKLZ	117

96.5

CA	BAKERSFIELD	KKXX	110
CA	SAN DIEGO	KYXY	114
CA	SAN FRANCISCO	KOIT	115
WA	SEATTLE-TACOM	KYCW	120

96.7

CA	CHICO	KZAP	110
CA	FRESNO	KEZL	110
CA	LOS ANGELES	KWIZ	111
CA	PALM SPRINGS	KUNA	113
CA	SAN BERNARD.	KCAL	114
WA	YAKIMA	KZTB	121

96.9

CA	MONTEREY-SALIN.	KWAV	112
CA	SACRAMENTO	KSEG	114
WA	YAKIMA	KZTA	121

97.1

CA	LOS ANGELES	KLSX	111
CA	MODESTO	KZMS	112
CA	SAN LUIS OBIS.	KWQH	116
CA	VISALIA	KSEQ	117
NV	LAS VEGAS	KXPT	117
OR	PORTLAND	KKSN	119

97.3

| CA | REDDING | KNCQ | 113 |

97.5 (right column)

CA	SAN DIEGO	KSON	115
CA	SAN FRANCISCO	KLLC	115
NV	RENO	KWNZ	118
WA	SEATTLE-TACOM	KBSG	120

97.5

| CA | MERCED | KABX | 112 |
| CA | SAN BERNARD. | KSSE | 114 |

97.7

CA	BAKERSFIELD	KRME	110
CA	CHICO	KZCO	110
CA	SAN FRANCISCO	KFFG	115
CA	SANTA ROSA	KMGG	117
CA	STOCKTON	KWIN	117

97.9

CA	FRESNO	KNAX	110
CA	LOS ANGELES	KLAX	111
CA	MONTEREY-SALIN.	KLXM	112
CA	SACRAMENTO	KTTA	114
OR	EUGENE	KKTT	118

98.1

CA	SAN DIEGO	KIFM	115
CA	SAN FRANCISCO	KISQ	115
CA	SAN LUIS OBIS.	KKJG	116
NV	RENO	KBUL	118
WA	SEATTLE-TACOM	KING	120
WA	SPOKANE	KISC	120

98.3

CA	LOS ANGELES	KRTO	111
CA	MODESTO	KWNN	112
CA	OXNARD-VENTUR.	KDAR	113
WA	RICH-PASCO-KEN	KEYW	120

98.5

CA	BAKERSFIELD	KSMJ	110
CA	PALM SPRINGS	KWXY	113
CA	SACRAMENTO	KRXQ	114
CA	SAN FRANCISCO	KUFX	115
NV	LAS VEGAS	KLUC	117

98.7

CA	LOS ANGELES	KYSR	111
CA	MERCED	KLOQ	112
OR	PORTLAND	KUPL	119

98.9

CA	SAN FRANCISCO	KSOL	115
CA	VISALIA	KSOF	117
WA	SEATTLE-TACOM	KWJZ	120
WA	SPOKANE	KKZX	120

99.1

CA	MONTEREY-SALIN.	KZOL	112
CA	SAN BERNARD.	KGGI	114
OR	EUGENE	KODZ	118

99.3

CA	BAKERSFIELD	KKBB	110
CA	FRESNO	KJWL	110
CA	REDDING	KNNN	113
CA	SAN DIEGO	XHKY	115
CA	STOCKTON	KJOY	117

99.5

CA	LOS ANGELES	KKLA	111
CA	MONTEREY-SALIN.	KLOK	112
OR	PORTLAND	KWJJ	119

99.7

CA	SAN FRANCISCO	KFRC	115
CA	SAN LUIS OBIS.	KWWV	116
CA	VISALIA	KIOO	117
WA	YAKIMA	KHHK	121

99.9

CA	MERCED	KCIV	112
CA	SAN BERNARD.	KOLA	114
CA	SANTA BARBARA	KTYD	116
WA	SEATTLE-TACOM	KISW	119
WA	SPOKANE	KXLY	120

100.1

CA	SANTA ROSA	KZST	117
CA	STOCKTON	KQOD	117
NV	RENO	KTHX	118

100.3

CA	LOS ANGELES	KCMG	111
CA	SAN FRANCISCO	KBRG	115
OR	MEDFORD	KRWQ	119
OR	PORTLAND	KKRZ	119

100.5

| CA | PALM SPRINGS | KPSI | 113 |
| CA | SACRAMENTO | KZZO | 114 |

F

CA	VISALIA	KOJJ	117
NV	LAS VEGAS	KMZQ	117

100.7

CA	MONTEREY-SALIN.	KTOM	112
CA	OXNARD-VENTUR.	KHAY	113
CA	SAN DIEGO	KFMB	114
CA	SAN FRANCISCO	KKHI	115
WA	SEATTLE-TACOM	KIRO	120

100.9

CA	STOCKTON	KMIX	117
NV	RENO	KRZQ	118
WA	YAKIMA	KARY	121

101.1

CA	FRESNO	KVSR	110
CA	LOS ANGELES	KRTH	111
CA	SACRAMENTO	KHYL	114
OR	PORTLAND	KUFO	119
WA	SPOKANE	KEYF	120

101.3

CA	SAN BERNARD.	KATY	114
CA	SAN FRANCISCO	KIOI	115
CA	SAN LUIS OBIS.	KSTT	116
WA	RICH-PASCO-KEN	KGDN	121

101.5

CA	BAKERSFIELD	KGFM	110
CA	CHICO	KMJE	110
CA	SAN DIEGO	KGB	114
WA	SEATTLE-TACOM	KPLZ	119

101.7

CA	MONTEREY-SALIN.	KXDC	112
CA	SAN BERNARD.	KXSB	114
CA	SAN FRANCISCO	KKIQ	115
CA	SANTA BARBARA	KSBL	116
CA	SANTA ROSA	KXFX	117
NV	RENO	KRNV	118
WA	RICH-PASCO-KEN	KZXR	120

101.9

CA	FRESNO	KOQO	110
CA	LOS ANGELES	KSCA	111
CA	SACRAMENTO	KRRE	114
NV	LAS VEGAS	KFMS	117
OR	MEDFORD	KCMX	118
OR	PORTLAND	KINK	119
WA	SPOKANE	KTSL	120

102.1

CA	MONTEREY-SALIN.	KRKC	112
CA	SAN DIEGO	KXST	114
CA	SAN FRANCISCO	KDFC	115

102.3

CA	LOS ANGELES	KJLH	111
CA	MODESTO	KJSN	112
CA	PALM SPRINGS	KJJZ	113

102.5

CA	BAKERSFIELD	KCNQ	110
CA	MONTEREY-SALIN.	KDON	112
CA	SACRAMENTO	KSFM	114
WA	SEATTLE-TACOM	KZOK	119

102.7

CA	FRESNO	KALZ	110
CA	LOS ANGELES	KIIS	111
WA	RICH-PASCO-KEN	KORD	120

102.9

CA	BAKERSFIELD	KSUV	110
CA	OXNARD-VENTUR.	KXLM	113
CA	SAN FRANCISCO	KBLX	115

103.1

CA	LOS ANGELES	KACD	111
CA	LOS ANGELES	KBCD	111
CA	PALM SPRINGS	KEZN	113
CA	SAN LUIS OBIS.	KBZX	116
WA	SPOKANE	KCDA	120

103.3

CA	MODESTO	KATM	112
CA	SANTA BARBARA	KRUZ	116
OR	PORTLAND	KKCW	119

103.5

CA	CHICO	KHSL	110
CA	LOS ANGELES	KOST	111
CA	MONTEREY-SALIN.	KRAY	112
CA	SACRAMENTO	KBMB	114
NV	LAS VEGAS	KISF	118
OR	MEDFORD	KOPE	119

103.7

CA	OXNARD-VENTUR.	KMLA	113
CA	SAN DIEGO	KPLN	114
CA	SAN FRANCISCO	KKSF	115
CA	VISALIA	KRZR	117
WA	SEATTLE-TACOM	KMTT	119

103.9

CA	BAKERSFIELD	KMYX	110
CA	LOS ANGELES	KACE	111
CA	MERCED	KDJK	112
CA	MONTEREY-SALIN.	KISE	112
CA	SAN BERNARD.	KCXX	114
WA	SPOKANE	KNJY	120

104.1

CA	MODESTO	KHKK	112
CA	SANTA ROSA	KMHX	117
CA	VISALIA	KFRR	117
WA	YAKIMA	KXDD	121

104.3

CA	BAKERSFIELD	KCOO	110
CA	BAKERSFIELD	KISV	110
CA	LOS ANGELES	KBIG	111
CA	MONTEREY-SALIN.	KMBY	112
CA	REDDING	KSHA	113
CA	SACRAMENTO	KQBR	114
NV	LAS VEGAS	KJUL	118

104.5

CA	BAKERSFIELD	KVLI	110
CA	SAN DIEGO	XLTN	115
CA	SAN FRANCISCO	KFOG	115
CA	SAN LUIS OBIS.	KIQO	116
NV	RENO	KDOT	118

104.7

CA	MERCED	KHTN	112
CA	OXNARD-VENTUR.	KCAQ	113
CA	PALM SPRINGS	KDES	113
OR	EUGENE	KDUK	118

104.9

CA	CHICO	KYIX	110
CA	SAN FRANCISCO	KLDZ	115
CA	SANTA ROSA	KRPQ	117
CA	VISALIA	KCRZ	117
WA	SEATTLE-TACOM	KKBY	120

105.1

CA	FRESNO	KLBN	110
CA	LOS ANGELES	KKGO	111
CA	MONTEREY-SALIN.	KOCN	112
CA	SACRAMENTO	KNCI	114
NV	LAS VEGAS	KVBC	118
OR	MEDFORD	KAKT	118
OR	PORTLAND	KRSK	119

105.3

CA	BAKERSFIELD	KKDJ	110
CA	SAN DIEGO	KIOZ	115
CA	SAN FRANCISCO	KITS	115
WA	RICH-PASCO-KEN	KONA	120
WA	SEATTLE-TACOM	KCMS	120

105.5

CA	LOS ANGELES	KBUE	111
CA	OXNARD-VENTUR.	KKBE	113
NV	LAS VEGAS	KSTJ	117

105.7

CA	SAN BERNARD.	KXRS	114
CA	SAN FRANCISCO	KARA	115
NV	RENO	KOZZ	118
WA	SPOKANE	KAEP	120
WA	YAKIMA	KRSE	121

105.9

CA	FRESNO	KRNC	110
CA	LOS ANGELES	KPWR	111

106.1

CA	BAKERSFIELD	KRAB	110
CA	PALM SPRINGS	KPLM	113
CA	REDDING	KRRX	113
CA	SAN FRANCISCO	KMEL	115
CA	SAN LUIS OBIS.	KWEZ	116
WA	SEATTLE-TACOM	KBKS	119

106.3

CA	LOS ANGELES	KGMX	111
CA	LOS ANGELES	KALI	111
CA	MERCED	KIBG	112
CA	MONTEREY-SALIN.	KLUE	112
CA	SANTA BARBARA	KHTY	116
CA	SANTA BARBARA	KMGQ	116

OR	MEDFORD	KZZE	118

106.5

CA	SACRAMENTO	KWOD	114
CA	SAN DIEGO	KKLQ	115
CA	SAN FRANCISCO	KEZR	115
NV	LAS VEGAS	KSNE	117
WA	RICH-PASCO-KEN	KEGX	120

106.5

CA	LOS ANGELES	KROQ	111
CA	VISALIA	KJUG	117
OR	PORTLAND	KKJZ	119

106.5

CA	MERCED	KQLB	112
NV	RENO	KRNO	118
OR	EUGENE	KCST	118
WA	SEATTLE-TACOM	KRWM	119

106.5

CA	BAKERSFIELD	KCWR	110
CA	LOS ANGELES	KLYY	111
CA	MONTEREY-SALIN.	KVRG	112
CA	OXNARD-VENTUR.	KVYY	113
CA	SAN DIEGO	KSYY	115

106.5

CA	SAN LUIS OBIS.	KQJZ	116
CA	STOCKTON	KSTN	117
WA	YAKIMA	KFFM	121

106.5

CA	LOS ANGELES	KLVE	111
CA	MONTEREY-SALIN.	KPIG	112
CA	VISALIA	KMPH	117
OR	MEDFORD	KKJJ	118
OR	PORTLAND	KBBT	119

106.5

CA	SAN FRANCISCO	KSAN	115
WA	SEATTLE-TACOM	KNDD	120

106.5

CA	BAKERSFIELD	KUZZ	110
CA	LOS ANGELES	KWVE	111
CA	SACRAMENTO	KDND	114

AM

540

CA	MONTEREY-SALIN.	KIEZ	113

550

CA	BAKERSFIELD	KUZZ	110

560

CA	SAN FRANCISCO	KSFO	116

570

CA	LOS ANGELES	KLAC	111
WA	SEATTLE-TACOM	KVI	120

580

CA	FRESNO	KMJ	110
OR	MEDFORD	KCMX	119

590

CA	SAN BERNARD.	KSZZ	114
OR	EUGENE	KUGN	118
WA	SPOKANE	KAQQ	120

600

CA	REDDING	KNRO	113
CA	SAN DIEGO	KOGO	115

610

CA	SAN FRANCISCO	KFRC	115
OR	MEDFORD	KRTA	119
WA	RICH-PASCO-KEN	KONA	120

620

OR	PORTLAND	KEWS	119

630

CA	MONTEREY-SALIN.	KIDD	112
NV	RENO	KPTT	118
WA	SEATTLE-TACOM	KCIS	120

640

CA	LOS ANGELES	KFI	111

650

CA	SACRAMENTO	KSTE	114

680

CA	SAN FRANCISCO	KNBR	116

700

CA	MONTEREY-SALIN.	KVRG	112

710

CA	LOS ANGELES	KDIS	111
CA	SACRAMENTO	KFIA	114

WA	SEATTLE-TACOM	KIRO	120

720

NV	LAS VEGAS	KDWN	118

740

CA	SAN FRANCISCO	KCBS	116

750

OR	PORTLAND	KXL	119

760

CA	SAN DIEGO	KFMB	115

770

CA	MODESTO	KCBC	112
WA	SEATTLE-TACOM	KNWX	120

780

NV	RENO	KKOH	118

790

CA	FRESNO	KOOR	110
CA	LOS ANGELES	KABC	111
WA	SPOKANE	KJRB	120

800

OR	PORTLAND	KPDQ	119

810

CA	SAN FRANCISCO	KGO	116

820

WA	SEATTLE-TACOM	KGNW	120

840

NV	LAS VEGAS	KXNF	118
OR	EUGENE	KKNX	118

850

WA	SEATTLE-TACOM	KHHO	120

860

CA	MODESTO	KTRB	112

870

CA	LOS ANGELES	KIEV	111
NV	LAS VEGAS	KLSQ	118
WA	RICH-PASCO-KEN	KFLD	121

880

CA	MONTEREY-SALIN.	KKMC	113
OR	MEDFORD	KTMT	119
WA	SEATTLE-TACOM	KIXI	120

900

CA	FRESNO	KBIF	110
CA	LOS ANGELES	KRRA	111

910

CA	OXNARD-VENTUR.	KOXR	113
CA	SAN FRANCISCO	KNEW	115
OR	PORTLAND	KFXX	119
OR	PORTLAND	KKSN	119

920

CA	MODESTO	KLOC	112
CA	PALM SPRINGS	KPSI	113
CA	SAN LUIS OBIS.	KVEC	116
NV	LAS VEGAS	KBAD	118
NV	RENO	KQLO	118
WA	SPOKANE	KXLY	120

930

CA	LOS ANGELES	KKHJ	111

940

CA	FRESNO	KFRE	111
OR	PORTLAND	KWBY	119

950

WA	SEATTLE-TACOM	KJR	120

960

CA	SAN FRANCISCO	KABL	115
WA	RICH-PASCO-KEN	KALE	121

970

CA	BAKERSFIELD	KAFY	110
CA	MODESTO	KANM	112
CA	PALM SPRINGS	KCLB	113
NV	LAS VEGAS	KNUU	118
OR	PORTLAND	KUPL	119
WA	SPOKANE	KTRW	120

980

CA	LOS ANGELES	KFWB	112
CA	MONTEREY-SALIN.	KCTY	112
WA	YAKIMA	KJOX	121

990

CA	SANTA BARBARA	KTMS	117

1000

CA	SAN DIEGO	KCEO	115
WA	SEATTLE-TACOM	KOMO	120

1010

CA	BAKERSFIELD	KCHJ	110
CA	PALM SPRINGS	KNWZ	113

F

CA	SAN FRANCISCO	KIQI	115

1020

CA	LOS ANGELES	KTNQ	111
WA	YAKIMA	KYXE	121

1030

CA	SAN LUIS OBIS.	KJDJ	116

1040

CA	SAN DIEGO	KURS	115

1050

CA	SAN FRANCISCO	KTCT	116
OR	EUGENE	KORE	118
WA	SPOKANE	KEYF	120

1070

CA	LOS ANGELES	KNX	112

1080

CA	MONTEREY-SALIN.	KSCO	113
OR	PORTLAND	KOTK	119
WA	SPOKANE	KVNI	120

1090

WA	SEATTLE-TACOM	KRPM	119

1100

CA	SAN FRANCISCO	KFAX	116

1110

CA	LOS ANGELES	KRLA	111

1120

OR	EUGENE	KPNW	118

1130

CA	SAN DIEGO	KSDO	115

1140

CA	PALM SPRINGS	KCMJ	113
CA	SACRAMENTO	KHTK	114

1150

CA	LOS ANGELES	KXTA	112

1170

CA	SAN DIEGO	KCBQ	115
CA	SAN FRANCISCO	KLOK	115

1180

CA	BAKERSFIELD	KERI	110

1190

OR	PORTLAND	KEX	119

1210

CA	SAN DIEGO	KPRZ	115
WA	SEATTLE-TACOM	KBSG	120
WA	YAKIMA	KREW	121

1220

CA	SAN FRANCISCO	KBPA	116

1230

CA	BAKERSFIELD	KGEO	110
CA	LOS ANGELES	KYPA	111
CA	SAN LUIS OBIS.	KPRL	116
NV	LAS VEGAS	KLAV	118
NV	RENO	KCBN	118
OR	MEDFORD	KSJK	119
OR	PORTLAND	KMUZ	119

1240

CA	MONTEREY-SALIN.	KNRY	113
CA	SACRAMENTO	KSQR	114
CA	SAN BERNARD.	KKLA	114
CA	SAN DIEGO	KSON	115
CA	VISALIA	KJOP	117
WA	SEATTLE-TACOM	KGY	119

1250

CA	SANTA BARBARA	KEYT	117

1260

CA	LOS ANGELES	KGIL	111
CA	SAN FRANCISCO	KOIT	115

1270

CA	PALM SPRINGS	KXPS	113
CA	VISALIA	KJUG	117
NV	RENO	KPLY	118

1280

CA	SAN LUIS OBIS.	KKAL	116
CA	STOCKTON	KJAX	117
WA	YAKIMA	KIT	121

1290

CA	CHICO	KPAY	110
CA	SAN BERNARD.	KKDD	114
CA	SANTA BARBARA	KZBN	116

1300

CA	FRESNO	KYNO	111
CA	LOS ANGELES	KAZN	111
NV	RENO	KPTL	118
WA	SEATTLE-TACOM	KKOL	120

1310

CA	SAN FRANCISCO	KMKY	116

1320

CA	SACRAMENTO	KCTC	114
CA	SAN DIEGO	KKSM	115

1330

CA	LOS ANGELES	KWKW	111
CA	MERCED	KLBS	112
OR	PORTLAND	KKPZ	119

1340

CA	FRESNO	KCBL	111
CA	MONTEREY-SALIN.	KOMY	113
CA	PALM SPRINGS	KWXY	113
CA	SAN LUIS OBIS.	KGLW	116
CA	SANTA BARBARA	KXXT	117
NV	LAS VEGAS	KRLV	118
NV	RENO	KXEQ	118
WA	RICH-PASCO-KEN	KTCR	121

1350

CA	SAN BERNARD.	KCKC	114
CA	SANTA ROSA	KSRO	117

1360

CA	MODESTO	KFIV	112
CA	SAN DIEGO	KPOP	115
OR	PORTLAND	KUIK	119

1370

CA	SAN BERNARD.	KWRM	114
CA	SAN FRANCISCO	KZSF	115
CA	VISALIA	KGEN	117

1380

CA	MONTEREY-SALIN.	KTOM	112
CA	SACRAMENTO	KTKZ	114

1390

CA	MODESTO	KVIN	112
WA	YAKIMA	KBBO	121

1400

CA	OXNARD-VENTUR.	KKZZ	113
CA	PALM SPRINGS	KESQ	113
CA	REDDING	KQMS	113
CA	SAN FRANCISCO	KVTO	116
CA	VISALIA	KFSO	117
OR	EUGENE	KNND	118

1410

CA	BAKERSFIELD	KERN	110
CA	SAN BERNARD.	KCAL	114

1420

CA	STOCKTON	KSTN	117
WA	SEATTLE-TACOM	KRIZ	120

1430

CA	FRESNO	KFIG	110
CA	SAN FRANCISCO	KVVN	116
OR	PORTLAND	KYKN	119

1440

CA	SAN BERNARD.	KDIF	114
OR	MEDFORD	KMED	119

1450

CA	OXNARD-VENTUR.	KVEN	113
CA	PALM SPRINGS	KGAM	113
CA	SAN DIEGO	KSPA	115
CA	SAN FRANCISCO	KEST	116
CA	VISALIA	KTIP	117
NV	RENO	KHIT	118
OR	EUGENE	KKXO	118

1460

CA	SANTA ROSA	KRRS	117
NV	LAS VEGAS	KENO	118
WA	YAKIMA	KMWX	121

1470

CA	SACRAMENTO	KOME	114

1480

CA	LOS ANGELES	KVNR	111
CA	MERCED	KYOS	112
OR	PORTLAND	KBMS	119

1490

CA	BAKERSFIELD	KWAC	110
CA	MONTEREY-SALIN.	KRKC	112
CA	SANTA BARBARA	KBKO	116
WA	YAKIMA	KENE	121

1500

CA	SAN FRANCISCO	KSJX	116

1510

WA	SPOKANE	KGA	120

F

CA	OXNARD-VENTUR.	KTRO	113
1530			
CA	SACRAMENTO	KFBK	114
1550			
CA	FRESNO	KXEX	110
CA	SAN FRANCISCO	KYCY	115
OR	PORTLAND	KVAN	119
1560			
CA	BAKERSFIELD	KNZR	110
1570			
CA	MONTEREY-SALIN.	KTGE	112
1580			
CA	MERCED	KTFN	112
1590			
CA	OXNARD-VENTUR.	KXFS	113
CA	SAN FRANCISCO	KLIV	116
1600			
CA	FRESNO	KGST	110

F

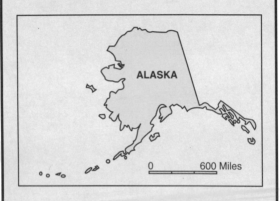

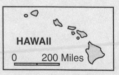

G

	FREQ	STATION	POWER	FORMAT	
	98.9	KYMG	IIIII	AC	
	103.1	KMXS	III	HOT AC	
	101.3	KGOT	III	TOP 40	
	100.5	KBFX	III	CLS RK	
	102.1	KKRO	III	CLS RK	
	106.5	KWHL	III	MOD RK	
	97.3	KEAG	IIIIIIII	OLDS	
	104.1	KBRJ	III	CTRY	
	107.5	KASH	III	CTRY	
	98.1	KLEF	III	CLS	
	105.3	KNIK	III	JZZ	
AM	590	KHAR	IIIIIIII	BB, NOS	
AM	1080	KASH	IIIII	CLS	
	88.1	KRUA	I	DVRS	C
	90.3	KNBA	IIIIIIII	ALT RK, NPR, PRI	C
	91.1	KSKA	IIIIIIII	DVRS, NPR, PRI	P
	89.3	KATB	III	RLG	P
AM	550	KTZN	IIIIIIII	SPRTS	
AM	650	KENI	IIIIIIII	SPRTS	
AM	700	KBYR	IIIIIIII	NWS, TLK	
AM	750	KFQD	IIIIIIII	NWS, TLK	
AM	1020	KAXX	IIIII	SPRTS	

FAIRBANKS, AK

	FREQ	STATION	POWER	FORMAT	
	101.1	KAKQ	III	AC	
	98.1	KWLF	III	TOP 40	
	95.9	KXLR	III	CLS RK	
	103.9	KUWL	I	DVRS	
	104.7	KKED	III	ROCK	
	102.5	KIAK	IIIIIIII	CTRY	
AM	1170	KJNP	IIIIIIII	CTRY	
	91.5	KSUA	I	MOD RK	C
	100.3	KJNP	IIIIIIII	RLG AC	
AM	660	KFAR	IIIIIIII	NWS, TLK	
AM	820	KCBF	IIIIIIII	NWS, TLK	
AM	970	KIAK	IIIIIIII	NWS, TLK	

JUNEAU, AK

	FREQ	STATION	POWER	FORMAT	
AM	800	KINY	IIIIIIII	AC	
	106.3	KSUP	III	CLS RK	
	105.1	KTKU	I	CTRY	
	104.3	KTOO	I	DVRS, NPR, PRI	P
AM	630	KJNO	IIIIIIII	NWS, TLK	

HILO, HI

	FREQ	STATION	POWER	FORMAT	
	93.9	KLUA	IIIIIIII	AC	
	94.7	KWXX	III	AC	
	95.9	KPVS	IIIIIIII	AC	
	97.9	KKBG	III	AC	
	106.1	KLEO	III	AC	
	106.9	KWYI	III	AC	
	97.1	KNWB	III	CLS RK	
	101.5	KAOY	III	CLS RK	
	92.7	KAOE	III	OLDS	
	100.3	KHWI	III	OLDS	
AM	790	KKON	IIIIIIII	OLDS	
AM	850	KHLO	IIIIIIII	OLDS	
AM	620	KIPA	IIIIIIII	BB, NOS	
AM	1060	KAHU	I	ETH	
	107.7	KKOA	III	URBAN	
AM	670	KPUA	IIIIIIII	NWS, TLK	

	FREQ	STATION	POWER	FORMAT	
	92.3	KSSK	IIIIIIII	AC	
	96.3	KRTR	IIIIIIII	AC	
AM	590	KSSK	IIIIIIII	AC	
	93.1	KQMQ	III	TOP 40	
	93.9	KIKI	III	TOP 40	
AM	690	KQMQ	IIIIIIII	TOP 40	
	97.5	KPOI	III	ROCK	
	98.5	KKLV	IIIIIIII	CLS RK	
	101.9	KUCD	IIIIIIII	DVRS	
	107.9	KGMZ	IIIIIIII	OLDS	
AM	1460	KULA	IIIII	OLDS	
	102.7	KHUL	IIIIIIII	CTRY	
AM	990	KIKI	IIIIIIII	CTRY	
AM	1500	KUMU	IIIII	BB, NOS	
	88.1	KHPR	IIIIIIII	CLS, NPR, PRI	P
	89.3	KIPO	III	NWS, TLK, NPR, PRI	P
	90.3	KTUH	I	DVRS	C
AM	1380	KIFO	IIIII	NWS, TLK, NPR, PRI	P
	94.7	KUMU	III	EZ	
	95.5	KAIM	IIIIIIII	RLG	
	100.3	KCCN	IIIIIIII	ETH	
	105.1	KINE	IIIIIIII	ETH	
AM	940	KJPN	IIIIIIII	ETH	
AM	1210	KZOO	I	ETH	
AM	1270	KNDI	IIIII	ETH	
AM	1420	KCCN	IIIII	ETH	
AM	650	KHNR	IIIIIIII	NWS	
AM	760	KGU	IIIIIIII	SPRTS	
AM	830	KHVH	IIIIIIII	NWS, TLK	

HI PR

SAN JUAN, PR

	FREQ	STATION	POWER	FORMAT	
	105.7	WCAD	IIIIIIII	CLS RK	
	91.3	WIPR	IIIIIIII	SPAN	
	94.7	WCOM	IIIIIIII	SPAN	
	98.5	WPRM	IIIIIIII	SPAN	
	99.9	WIOA	IIIII	SPAN	
	100.7	WXYX	IIIIIIII	SPAN	
	102.5	WIAC	IIIIIIII	SPAN	
	103.7	WCFI	IIIII	SPAN	
	104.7	WKAQ	IIIIIIII	SPAN	
AM	580	WKAQ	IIIIIIII	SPAN	
AM	630	WSKN	IIIIIIII	SPAN	
AM	680	WAPA	IIIIIIII	SPAN	
AM	740	WIAC	IIIIIIII	SPAN	
AM	810	WKVM	IIIIIIII	SPAN	
AM	870	WQBS	IIIIIIII	SPAN	
AM	940	WIPR	IIIIIIII	SPAN	P
AM	1090	WSOL	I	SPAN	
AM	1410	WRSS	I	SPAN	
AM	1460	WLRP	I	SPAN	
AM	1520	WVOZ	IIIII	SPAN	
AM	1560	WRSJ	IIIII	SPAN	
AM	1030	WOSO	IIIII	BB, NOS	
AM	1600	WLUZ	IIIII	BB, NOS	
	89.7	WRTU	IIIIIIII	DVRS, PRI	C

88.1			
AK	ANCHORAGE	KRUA	129
HI	HONOLULU	KHPR	130
88.3			
AK	ANCHORAGE	KATB	129
HI	HONOLULU	KIPO	130
88.7			
PR	SAN JUAN	WRTU	130
90.3			
AK	ANCHORAGE	KNBA	129
HI	HONOLULU	KTUH	130
91.1			
AK	ANCHORAGE	KSKA	129
91.3			
PR	SAN JUAN	WIPR	130
91.5			
AK	FAIRBANKS	KSUA	129
92.3			
HI	HONOLULU	KSSK	130
92.7			
HI	HILO	KAOE	129
93.1			
HI	HONOLULU	KQMQ	130
93.9			
HI	HILO	KLUA	129
HI	HONOLULU	KIKI	130
94.7			
HI	HILO	KWXX	129
HI	HONOLULU	KUMU	130
PR	SAN JUAN	WCOM	130
95.5			
HI	HONOLULU	KAIM	130
95.9			
AK	FAIRBANKS	KXLR	129
HI	HILO	KPVS	129
96.3			
HI	HONOLULU	KRTR	130
97.1			
HI	HILO	KNWB	129
97.3			
AK	ANCHORAGE	KEAG	129
97.5			
HI	HONOLULU	KPOI	130
97.9			
HI	HILO	KKBG	129
98.1			
AK	ANCHORAGE	KLEF	129
AK	FAIRBANKS	KWLF	129
98.5			
HI	HONOLULU	KKLV	130
PR	SAN JUAN	WPRM	130
98.9			
AK	ANCHORAGE	KYMG	129
99.9			
PR	SAN JUAN	WIOA	130
100.3			
AK	FAIRBANKS	KJNP	129
HI	HILO	KHWI	129
HI	HONOLULU	KCCN	130
100.5			
AK	ANCHORAGE	KBFX	129
100.7			
PR	SAN JUAN	WXYX	130
101.1			
AK	FAIRBANKS	KAKQ	129
101.3			
AK	ANCHORAGE	KGOT	129
101.5			
HI	HILO	KAOY	129
101.9			
HI	HONOLULU	KUCD	130
102.1			
AK	ANCHORAGE	KKRO	129
102.5			
AK	FAIRBANKS	KIAK	129
PR	SAN JUAN	WIAC	130
102.7			
HI	HONOLULU	KHUL	130
103.1			
AK	ANCHORAGE	KMXS	129
103.7			
PR	SAN JUAN	WCFI	130
103.9			
AK	FAIRBANKS	KUWL	129

104.1			
AK	ANCHORAGE	KBRJ	129
104.3			
AK	JUNEAU	KTOO	129
104.7			
AK	FAIRBANKS	KKED	129
PR	SAN JUAN	WKAQ	130
105.1			
AK	JUNEAU	KTKU	129
HI	HONOLULU	KINE	130
105.3			
AK	ANCHORAGE	KNIK	129
105.7			
PR	SAN JUAN	WCAD	130
106.1			
HI	HILO	KLEO	129
106.5			
AK	JUNEAU	KSUP	129
106.7			
AK	ANCHORAGE	KWHL	129
106.9			
HI	HILO	KWYI	129
107.5			
AK	ANCHORAGE	KASH	129
107.7			
HI	HILO	KKOA	129
107.9			
HI	HONOLULU	KGMZ	130

AM			
550			
AK	ANCHORAGE	KTZN	129
580			
PR	SAN JUAN	WKAQ	130
590			
AK	ANCHORAGE	KHAR	129
HI	HONOLULU	KSSK	130
620			
HI	HILO	KIPA	129
630			
AK	JUNEAU	KJNO	129
PR	SAN JUAN	WSKN	130
650			
AK	ANCHORAGE	KENI	129
HI	HONOLULU	KHNR	130
660			
AK	FAIRBANKS	KFAR	129
670			
HI	HILO	KPUA	129
680			
PR	SAN JUAN	WAPA	130
690			
HI	HONOLULU	KQMQ	130
700			
AK	ANCHORAGE	KBYR	129
740			
PR	SAN JUAN	WIAC	130
750			
AK	ANCHORAGE	KFQD	129
760			
HI	HONOLULU	KGU	130
790			
HI	HILO	KKON	129
800			
AK	JUNEAU	KINY	129
810			
PR	SAN JUAN	WKVM	130
820			
AK	FAIRBANKS	KCBF	129
830			
HI	HONOLULU	KHVH	130
850			
HI	HILO	KHLO	129
870			
PR	SAN JUAN	WQBS	130
940			
HI	HONOLULU	KJPN	130
PR	SAN JUAN	WIPR	130
970			
AK	FAIRBANKS	KIAK	129

G

	990		
HI	HONOLULU	KIKI	130
	1020		
AK	ANCHORAGE	KAXX	129
	1030		
PR	SAN JUAN	WOSO	130
	1060		
HI	HILO	KAHU	129
	1080		
AK	ANCHORAGE	KASH	129
HI	HONOLULU	KWAI	130
	1090		
PR	SAN JUAN	WSOL	130
	1170		
AK	FAIRBANKS	KJNP	129
	1190		
PR	SAN JUAN	WBMJ	130
	1210		
HI	HONOLULU	KZOO	130
	1270		
HI	HONOLULU	KNDI	130
	1380		
HI	HONOLULU	KIFO	130
	1410		
PR	SAN JUAN	WRSS	130
	1420		
HI	HONOLULU	KCCN	130
	1460		
HI	HONOLULU	KULA	130
PR	SAN JUAN	WLRP	130
	1500		
HI	HONOLULU	KUMU	130
	1520		
PR	SAN JUAN	WVOZ	130
	1560		
PR	SAN JUAN	WRSJ	130
	1600		
PR	SAN JUAN	WLUZ	130

G

ICE HOCKEY

BOSTON	BRUINS		1030	WBZ
BUFFALO	SABRES		550	WGR
CALGARY	FLAMES		660	CFFR
CHICAGO	BLACK HAWKS		1000	WMVP
DALLAS	STARS		820	WBAP
DENVER	AVALANCHE		950	KKFN
DETROIT	RED WINGS		760	WJR
EDMONTON	OILERS		630	CHED
HARTFORD	WHALERS		1080	WTIC
ANAHEIM	MIGHTY DUCKS		95.9	KEZY
LOS ANGELES	KINGS		690	XTRA
MIAMI	PANTHERS		560	WQAM
MONTREAL	CANADIENS		800	CJAD
NEW YORK	DEVILS		770	WABC
NEW YORK	ISLANDERS	FM	98.5	WLIR
NEW YORK	RANGERS		660	WFAN
OTTAWA	SENATORS		580	CFRA
PHILADELPHIA	FLYERS		610	WIP
PHOENIX	COYOTES	FM	97.9	KUPD
PITTSBURGH	PENGUINS		1250	WTAE
SAN JOSE	SHARKS		610	KFRC
ST. LOUIS	BLUES		1120	KMOX
TAMPA / ST. PETE	LIGHTNING		910	WFNS
TORONTO	MAPLE LEAFS		640	CHOG
VANCOUVER	CANUCKS		980	CKNW
WASHINGTON D.C.	CAPITOLS		570	WTEM

BASEBALL

ATLANTA	BRAVES		750	WSB	
BALTIMORE	ORIOLES		1090	WBAL	
BOSTON	RED SOX		680	WRKO	
CHICAGO	CUBS		720	WGN	
CHICAGO	WHITE SOX		560	WIND	SPAN
CHICAGO	WHITE SOX		670	WMAQ	
CINCINNATI	REDS		700	WLW	
CLEVELAND	INDIANS		1220	WKNR	
DALLAS	RANGERS		910	KXEB	SPAN
DALLAS	RANGERS		1080	KRLD	
DENVER	ROCKIES	FM	103.5	KRFX	
DENVER	ROCKIES		850	KOA	
DETROIT	TIGERS		760	WJR	
HOUSTON	ASTROS		950	KPRC	
HOUSTON	ASTROS		1320	KXYZ	SPAN
KANSAS CITY	ROYALS	FM	97.3	WIBW	
LOS ANGELES	ANGELS		790	KABC	
LOS ANGELES	DODGERS		790	KABC	
LOS ANGELES	DODGERS		1330	KWKW	SPAN
MIAMI	MARLINS		560	WQAM	
MIAMI	MARLINS		1210	WCMQ	SPAN
MILWAUKEE	BREWERS		620	WTMJ	
MINN. - ST. PAUL	TWINS		830	WCCO	
MONTREAL	EXPOS		600	CIQC	
MONTREAL	EXPOS		730	CKAC	FREN
NEW YORK	METS	FM	97.9	WSKQ	SPAN
NEW YORK	METS		660	WFAN	
NEW YORK	YANKEES		770	WABC	
OAKLAND	ATHLETICS	FM	99.7	KFRC	
OAKLAND	ATHLETICS		610	KFRC	
OAKLAND	ATHLETICS		1430	KNTA	SPAN
PHILADELPHIA	PHILLIES	FM	98.1	WOGL	
PITTSBURGH	PIRATES		1020	KDKA	
SAN DIEGO	PADRES	FM	100.7	KFMB	
SAN DIEGO	PADRES		1420	XEXX	SPAN
SAN FRANCISCO	GIANTS		680	KNBR	
SAN FRANCISCO	GIANTS		1010	KIQI	SPAN
SEATTLE	MARINERS		710	KIRO	
ST. LOUIS	CARDINALS		1120	KMOX	
TORONTO	BLUE JAYS		590	CJCL	

SECTION H: SPORTS TEAMS
FOOTBALL • BASKETBALL

FOOTBALL

City	Team	Band	Freq	Call	Note
ATLANTA	FALCONS	FM	92.9	WZGC	
BALTIMORE	RAVENS		1300	WJFK	
BOSTON	PATRIOTS	FM	104.1	WBCN	
BUFFALO	BILLS		930	WBEN	
CHARLOTTE	PANTHERS		1110	WBT	
CHICAGO	BEARS		720	WGN	
CINCINNATI	BENGALS		550	WCKY	
DALLAS	COWBOYS	FM	103.7	KVIL	
DENVER	BRONCOS		850	KOA	
DETROIT	LIONS		1270	WXYT	
GREEN BAY	PACKERS		620	WTMJ	
HOUSTON	OILERS		740	KTRH	
INDIANAPOLIS	COLTS		1070	WIBC	
JACKSONVILLE	JAGUARS		690	WOKV	
KANSAS CITY	CHIEFS	FM	101.1	KCFX	
MIAMI	DOLPHINS		610	WIOD	
MIAMI	DOLPHINS		1210	WCMQ	SPAN
MINN. - ST. PAUL	VIKINGS	FM	102.1	KEEY	
MINN. - ST. PAUL	VIKINGS		1130	KFAN	
NEW ORLEANS	SAINTS		870	WWL	
NEW YORK	GIANTS		710	WOR	
NEW YORK	JETS		660	WFAN	
OAKLAND	RAIDERS	FM	93.3	KYCY	
OAKLAND	RAIDERS		1170	KLOK	SPAN
PHILADELPHIA	EAGLES	FM	94.1	WYSP	
PHOENIX	CARDINALS	FM	96.9	KHTC	
PHOENIX	CARDINALS		860	KUVA	SPAN
PITTSBURGH	STEELERS		1250	WTAE	
SAN DIEGO	CHARGERS		690	XTRA	
SAN FRANCISCO	49ers		810	KGO	
SEATTLE	SEA HAWKS		710	KIRO	
ST. LOUIS	RAMS	FM	93.7	KSD	
ST. LOUIS	RAMS		550	KSD	
TAMPA / ST. PETE	BUCCANEERS	FM	99.5	WQYK	
WASHINGTON D.C.	REDSKINS	FM	106.7	WJFK	

H

BASKETBALL

City	Team	Freq	Call	Note
ATLANTA	HAWKS	750	WSB	
BOSTON	CELTICS	850	WEEI	
CHARLOTTE	HORNETS	1110	WBT	
CHICAGO	BULLS	1000	WMVP	
CLEVELAND	CAVALIERS	1100	WTAM	
DALLAS	MAVERICKS	570	KLIF	
DENVER	NUGGETS	950	KKFN	
DETROIT	PISTONS	950	WWJ	
HOUSTON	ROCKETS	740	KTRH	
INDIANAPOLIS	PACERS	1070	WIBC	
LOS ANGELES	CLIPPERS	690	XTRA	
LOS ANGELES	LAKERS	570	KLAC	
LOS ANGELES	LAKERS	1330	KWKW	SPAN
MIAMI	HEAT	610	WIOD	
MILWAUKEE	BUCKS	620	WTMJ	
MINN. - ST. PAUL	TIMBER WOLVES	1130	KFAN	
NEW JERSEY	NETS	710	WOR	
NEW YORK	KNICKS	660	WFAN	
ORLANDO	MAGIC	580	WDBO	
PHILADELPHIA	76ers	610	WIP	
PHOENIX	SUNS	620	KTAR	
PORTLAND	TRAIL BLAZERS	1190	KEX	
SACRAMENTO	KINGS	1140	KHTK	
SALT LAKE CITY	JAZZ	1320	KFNZ	
SAN ANTONIO	SPURS	760	KTKR	
SAN ANTONIO	SPURS	1200	WOAI	
SAN FRANCISCO	WARRIORS	680	KNBR	
SEATTLE	SUPERSONICS	950	KJR	
TORONTO	RAPTORS	1010	CFRB	
VANCOUVER	GRIZZLIES	980	CKNW	
WASHINGTON D.C.	BULLETS	980	WWRC	

AIRLINES

Air Canada	800.776.3000
Air France	800.237.2747
Air Tran	800.247.8726
Alaska	800.426.0333
Alitalia	800.223.5730
All Nippon	800.235.9262
America West	800.235.9292
American	800.433.7300
British Airways	800.247.9297
Canadian	800.426.7000
Continental	800.525.0280
Delta	800.221.1212
Japan Airlines	800.525.3663
Kiwi International	800.538.5494
KLM	800.374.7747
Lufthansa	800.645.3880
Midwest Express	800.452.2022
Northwest	800.225.2525
Southwest	800.435.9792
TWA	800.221.2000
United	800.241.6522
US Airways	800.428.4322
Virgin Atlantic	800.862.8621

HOTELS

Best Western	800.528.1234
Choice Hotels	800.221.2222
Courtyard	800.321.2211
Days Inn	800.235.2525
Double Tree	800.528.0444
Embassy Suites	800.362.2779
Four Seasons	800.332.3442
Hampton Inn	800.426.7866
Hilton	800.445.8667
Holiday Inn	800.465.4329
Howard Johnson	800.654.2000
Hyatt	800.228.9000
Intercontinental	800.327.0200
Marriott	800.228.9290
Motel 6	800.466.8356
Omni	800.843.6664
Quality Inns	800.228.5151
Radisson	800.333.3333
Ramada	800.228.2828
Residence Inn	800.331.3131
Sheraton	800.325.3535
Stouffer Renaissance	800.468.3571
Westin	800.228.3000
Wyndham	800.822.4200

CAR RENTALS

Alamo	800.327.9633
Avis	800.331.1212
Budget	800.527.0700
Dollar	800.800.4000
Enterprise	800.325.8007
Hertz	800.654.3131
National	800.227.7368
Thrifty	800.367.2277

Essential RADIO GUIDE ®

For information on customized editions OF ESSENTIAL RADIO GUIDE ® or bulk orders of the retail edition, please contact PEREGRINE PRESS at 781.639.8090. National, Regional and Format specific editions of ESSENTIAL RADIO GUIDE are available, as well as Market specific ESSENTIAL RADIO CARDS ®. Embossed covers are also available. For larger orders, additional sections containing company or organization specific information can be included.

For comprehensive database and demographic information on Radio and Television stations in the United States and Radio stations in Canada, please contact THE CENTER FOR RADIO INFORMATION (CRI). CRI provides a broad range of resources to the broadcast industry to include custom databases, call letter availability, database audits, Arbitron® analyses, mailing lists, phone lists, broadcast fax services and other marketing and demographic information services

THE CENTER FOR RADIO INFORMATION
19 Market Street
Cold Spring, NY 10516

Tel: 800.359.9898   Fax: 914.265.2715

E-mail: **info@the-cri.com**
Web Site: **www.the-cri.com**

PEREGRINE PRESS LTD.
P.O. Box 363
Marblehead, MA 01945

Tel: 781.639.8090
Fax: 781.631.6341

E-mail: info@webradionet.com

Coming soon to a modem near you . . .
www.webradionet.com